THE HUMANE METROPOLIS

THE HUMANE METROPOLIS

People and Nature in the 21st-Century City

Edited by Rutherford H. Platt

University of Massachusetts Press · Amherst & Boston
in association with
Lincoln Institute of Land Policy · Cambridge

Copyright © 2006 by Rutherford H. Platt and the Lincoln Institute of Land Policy
All rights reserved
Printed in the United States of America

LC 2006019435
ISBN 1-55849-553-3 (library cloth ed.); 554-1 (paper)

Set in Adobe Minion and ITC Franklin Gothic by Dix
Printed and bound by The Maple-Vail Book Manufacturing Group

Library of Congress Cataloging-in-Publication Data

The humane metropolis : people and nature in the twenty-first-century city / edited by
Rutherford H. Platt.
    p. cm.
Conference papers.
Includes bibliographical references and index.
ISBN 1-55849-554-1 (pbk. : alk. paper)—ISBN 1-55849-553-3 (library cloth)
1. Urban ecology—United States—Congresses.   2. City planning—United States—Congresses.
3. Greenbelts—United States—Congresses.
I. Platt, Rutherford H.   II. Lincoln Institute of Land Policy.
HT243.U6H86 2006
307.760973—dc22
                    2006019435

British Library Cataloguing in Publication data are available.

To the editor's father,

RUTHERFORD PLATT (1894–1975)

Author, Photographer, Naturalist

This winter morning I took an hour's walk in Gramercy Park in the heart of New York. Some fifty well-groomed trees are the asset of this park for the shade they cast in summer, and for raising rents when apartment windows look out on them. But I have never seen anybody look at the trees in winter; they receive no more attention than black dead sticks. People on the benches were working the cross-word puzzles in the Sunday paper. In an atmosphere of so much indifference, one feels a little foolish staring at the trees and reaching for a twig to pull it down and examine the end buds.

But it makes a good hour's diversion. The clues were all there—just as they are out in the country. The silver tam-o'-shanters of the dogwood buds; the long varnished pyramids of the poplar; the bright red tridents of the red maple; the fat buds of the magnolia as furry as a cat's paw; and the crumpled black wells of the locust in which the buds are hidden. One by one I told off their names, and checked the answer with the bark and twigs and branches. It's marvelous how the potencies packed in a tiny seed had imprinted each vast structure with the clear-cut resemblances of its kind, and equipped it perfectly to the minutest detail.

From *This Green World* (Dodd Mead first printing, 1942; awarded John Burroughs Medal, 1945)

Contents

Acknowledgments

Many individuals and organizations have contributed to the achievement of this book, as well as the Humane Metropolis film, and the 2002 conference from which both were derived. First, I thank Armando Carbonell and the Lincoln Institute of Land Policy for supporting the conference and providing additional financial assistance as copublisher of this book. I also thank the Wyomissing Foundation and the late Laurance S. Rockefeller for their respective contributions to the conference and the film.

The chapter authors have been wonderful. Over the long publication process, they have remained enthusiastic, responsive, and patient. I truly appreciate their participation in the project, one and all.

Next, I want to thank Bruce Wilcox, Clark Dougan, Carol Betsch, Jack Harrison, and other members of the superb University of Massachusetts Press staff. It has been a real pleasure to collaborate with UMass Press on this sequel to our 1994 book *The Ecological City* (which remains in print).

Kathleen M. Lafferty of Roaring Mountain Editorial Services, as copy editor and technical advisor throughout the publication process, was infinitely helpful in creating a book out of this multiauthor (and multiego!) collection of manuscripts. Kathleen attended the original conference, where she met many of the authors, and has interacted with them throughout the production process. In this our third book together, she continues to calmly pilot the ship through the rocks and shoals of editorial chaos.

Finally, I am also very grateful to three graduate student collaborators, Amanda Krchak, Ted White, and Laurin N. Sievert, for their respective roles in the project. Amanda facilitated the flow of manuscripts between our office, the authors, and UMass Press; Ted created the twenty-two-minute *Humane Metropolis* film; and Laurin has been indispensable throughout the entire project as conference organizer, advisor on the film, technical guru for the book manuscripts, and chapter author.

Rutherford H. Platt

THE HUMANE METROPOLIS

Introduction

Humanizing the Exploding Metropolis

Rutherford H. Platt

A Subversive Little Book

"This is a book by people who like cities." Thus began William H. Whyte Jr.'s introduction to a subversive little book with the polemical title *The Exploding Metropolis: A Study of the Assault on Urbanism and How Our Cities Can Resist It* (Editors of *Fortune* 1957, hereinafter cited as TEM). Drawing on a roundtable of urban experts convened by two prominent magazines, *Fortune* and *Architectural Forum,* the book in six short essays reexamined the nature of cities and city building in the postwar era. The book also defined future agendas for "Holly" Whyte (as he was fondly known by his friends) and fellow editor Jane Jacobs.

Whyte had recently gained fame for his sociological critique of postwar business culture and suburban lifestyle, *The Organization Man* (Whyte 1956/2002). His 1957 *Exploding Metropolis* essay titled "Urban Sprawl" (perhaps the first use of that term) deplored the senseless loss of farmland and rural amenities due to suburban development, a theme expanded in his later book *The Last Landscape* (Whyte 1968/ 2002). Similarly, Jane Jacobs's essay "Downtown Is for People," which challenged conventional wisdom on urban renewal, foreshadowed her 1961 classic, *The Death and Life of Great American Cities.*

It was somewhat ironic for *Fortune*'s editors to take the lead in condemning postwar urban renewal and urban sprawl as that magazine had helped foster both. According to environmental historian Adam Rome (2001, 34–35), *Fortune* "published dozens of articles in 1946 and 1947 on the housing shortage. In a rare editorial—'Let's Have Ourselves a Housing Industry'—the editors supported a handful of government initiatives to encourage builders to operate on a larger scale." They even called for construction of public housing by government agencies, arguing that "if the government acted prudently to strengthen the housing market . . . the result would be the best defense against socialism, not a defeat for free enterprise" (Rome 2001, 35).

Congress rose to the challenge. In the late 1940s and early 1950s, it created a variety of new housing stimulus programs under the aegis of the Federal Housing Authority and the Veterans Administration. These programs helped fuel a construction boom of some fifteen million new housing units during the 1950s. Exemplars of these new white middle-class postwar suburbs were the two Levittowns in New York and Pennsylvania, and Park Forest, Illinois (the home of Whyte's archetypal "Organization Man"). The expansion of white suburbia was further subsidized by the federal interstate highway system authorized by Congress in 1956 and by federal tax deductions for mortgage interest, local property taxes, and accelerated depreciation for commercial real estate investments (Platt 2004, ch. 6).

For those left behind in the central cities, Congress established the federal urban renewal program in the housing acts of 1949 and 1954 to clear and redevelop "blighted areas." The standard model for redevelopment was the high-rise public or subsidized apartment project loosely modeled on the French architect Le Corbusier's ideal town plan, *La Ville Radieuse*—decried by Whyte as "the wrong design in the wrong place at the wrong time" (TEM 1957, xi). Such projects offered rental but not ownership units. Occupants were thus ineligible for federal home ownership tax deductions, assuming they had income against which to claim deductions, and also lacked the opportunity to build equity in the rising value of an owned home. The best of these apartment complexes, such as Metropolitan Life's Stuyvesant Town and Peter Cooper Village in Manhattan, were privately sponsored with government assistance. The worst, such as the infamous Pruitt-Igoe project in St. Louis and the Robert Taylor Homes in Chicago, both now demolished, were built by public housing authorities.

The Exploding Metropolis challenged both suburban and central city postwar construction on aesthetic and functional considerations. Prevailing patterns of land development on the urban fringe were ugly and inefficient, while redevelopment in the urban core was ugly and unsafe. Concerning the fringe, Whyte laments:

> Aesthetically, the result is a mess. It takes remarkably little blight to color a whole area; let the reader travel along a stretch of road he is fond of, and he will notice how a small portion of open land has given amenity to the area. But it takes only a few badly designed developments or billboards or hot-dog stands to ruin it, and though only a little bit of the land is used, the place will *look* filled up.
>
> Sprawl is bad esthetics; it is bad economics. Five acres are being made to do the work of one, and do it very poorly. This is bad for the farmers, it is bad for communities, it is bad for industry, it is bad for utilities, it is bad for the railroads, it is bad for the recreation groups, it is bad even for the developers. (TEM, 116–17)

And concerning central city housing:

> The scale of the projects is uncongenial to the human being. The use of the open space is revealing; usually it consists of manicured green areas carefully chained off lest they be profaned, and sometimes, in addition, a big central mall so vast and abstract as to be vaguely oppressive. There is nothing close for the eye to light on, no sense of intimacy or of things being on a human scale. (TEM, 21)

Concern with the visual appearance of urban places, of course, did not begin or end with Holly Whyte. Since the City Beautiful movement at the turn of the twentieth century, urban aesthetics had been a prevalent concern of architects and urbanists. Later critics of the visual urban landscape included Peter Blake, Donald Appleyard, Kevin Lynch, Ian McHarg, Tony Hiss, and James Howard Kunstler. Few, however, have articulated the nexus between urban form and function in simpler, more direct terms than Whyte.

The Exploding Metropolis, though, was sadly deficient in recognizing the social injustice of urban sprawl, namely the preferential treatment of the white middle class over the nonwhite poor in federal housing and tax policies, as well as the use of exclusionary zoning by suburban communities. As historian Kenneth T. Jackson documented in his 1985 book *Crabgrass Frontier*, federal housing authorities practiced "redlining" of neighborhoods by race and income to ensure that most new units built with federal assistance were suburban single-family homes for the white middle class. Even Donald Seligman's essay "The Enduring Slums" in *The Exploding Metropolis* blandly observed that "the white urban culture they [poor nonwhites] might assimilate *into* is receding before them; it is drifting off into the suburbs" (TEM 1957, 97). "Drifting off" is certainly a nonjudgmental way to describe the process of white flight in response to the *pull* of government incentives for suburban development and the reciprocal *push* of central city neglect. (Whyte in fact acknowledged that federal housing subsidies benefit "high-income people" in suburbia, whereas public housing programs benefit the poor in the cities, leading to a curious suggestion that the "middle class" also should be subsidized—to stay in the city [TEM 1957, 6]!)

During the 1950s, the central cities of the twenty largest metropolitan areas gained only 0.1 percent in population, whereas their suburbs grew by 45 percent (Teaford 1993, 98). Whether people "liked cities" or not was often secondary to whether they would pay the economic and emotional price of staying in them (especially if they had children) rather than fleeing to what a *New Yorker* magazine cover cartoon of December 10, 2001, slyly termed "Outer Perturbia." Obviously, most chose the latter, whether out of choice or necessity. National policies tilted in that direction and further polarized the metropolis between haves and have-nots.

This myopia concerning race, poverty, and the underlying dynamics of urban sprawl was by no means limited to *The Exploding Metropolis*. With the exception of the early "muckraker" urban reformers like Jacob Riis (1890), most urban scholarship before the 1960s had focused on economics and technology, not social equity. Even the literature on "human ecology" by progressive urban sociologists at the University of Chicago in the 1920s complacently referred to "so-called 'slums' and 'badlands,' with their submerged regions of poverty, degradation, and disease, and their underworlds of crime and vice" and a truly racist flourish: "Wedging out from here is the Black Belt, with its free and disorderly life" (Burgess 1925, 54–56). As late as 1961, French geographer Jean Gottmann in his classic *Megalopolis* effusively described the northeast urban corridor from Boston to Washington, D.C., as "a stupendous monument erected by titanic efforts" (Gottmann 1961, 23). Concerning poverty, however, he laconically wrote that "the labor market of the great cities still attracts large numbers of in-migrants from the poorer sections . . . especially Southern Negroes and Puerto Ricans, who congregate in the old urban areas and often live in slums" (Gottmann 1961, 66).

The Exploding Metropolis, however, was indeed revolutionary for its day in at least four respects. First, it rejected the conventional wisdom that suburbs are necessarily preferable to "real cities." Second, it urged that cities should be thought of, in effect, as habitats for people, not simply as centers of economic production, transportation nodes, or grandiose architectural stage sets. Third, it challenged the prevailing notion that population density ("crowding") is necessarily bad. Fourth, it established a precedent for more searching critiques of urban policies and programs in the coming decades, including but by no means limited to those of Holly Whyte and Jane Jacobs themselves. It marked the emergence of the nontechnician as self-taught "urban expert" and the rediscovery of the city as a "place," not just a complex of systems. In short, *The Exploding Metropolis* fired an early salvo of the debate over the nature, purpose, and design of city space that continues to rage today.

"The Observation Man"

On January 2, 2000, the *New York Times Magazine* in a series "People of the Millennium" profiled William H. Whyte (1917–99) as "The Observation Man" (a descriptor earlier applied to him by planner Eugenie Birch in 1986; see her essay in this volume). Norman Glazer (1999, 27) characterized him in the *Wilson Quarterly* as "the man who loved cities . . . one of America's most influential observers of the city and the space around it." Brendan Gill (1999, 99) in the *New Yorker* placed Whyte in company with other "learned amateurs"—Frederick Law Olmsted, Lewis Mumford, and Jane Jacobs—who became "our leading authorities on the nature of cities." Posing a series of questions about cities, Gill wrote: "The person best fitted to answer these questions is himself a seasoned New Yorker . . . who has been subjecting the city to a scrutiny as close as that to which Thoreau—still another learned amateur—subjected Walden Pond and its environs. His name is William H. Whyte and his equivalent to Thoreau's cabin is a narrow, high-stooped brownstone in the East Nineties."

A native of the picturesque Brandywine Valley in eastern Pennsylvania, Holly Whyte graduated from Princeton in 1939 and fought at Guadalcanal as an officer in the U.S. Marine Corps (figure 1). As discussed in the next essay, he joined the editorial staff of *Fortune* in New York after the war and began to examine the culture and habitats of postwar suburbia. In part 7 of his 1956 book *The Organization Man,* "The New Suburbia . . . ," Whyte analyzed the social geography of young corporate families living in the planned postwar suburb of Park Forest, Illinois.

Figure 1 William H. Whyte in the mid-1950s. (Photo courtesy of Alexandra Whyte.)

Literally mapping the patterns of social activities, parent-teacher association meetings, bridge games, and such, he determined that social interaction is promoted or inhibited by the spatial layout of homes, parking, yards, and common spaces, which in turn influenced the formation of friendships versus social isolation. He thus began a lifetime devoted to understanding better how the design of common or public spaces (e.g., parks, sidewalks, plazas) affects the lives and well-being of people who share them. This theme would later be further explored in *The Social Life of Small Urban Spaces* (Whyte 1980) and his capstone book, *City: Rediscovering the Center* (Whyte 1988).

Whyte left *Fortune* in 1959 to pursue a broader array of urban projects. His first technical publication on conservation easements (Whyte 1959) became the model for open space statutes in California, New York, Connecticut, Massachusetts, and Maryland (Birch 1986). As a consultant to the congressionally chartered Outdoor Recreation Resources Review Commission, he wrote a report on "Open Space Action" (Whyte 1962). The commission chair, Laurance S. Rockefeller, would support Whyte's work on urban land problems with a salary and an office in Rockefeller

Center for the rest of his career (Winks 1997). Whyte served as a member of President Lyndon Johnson's Task Force on Natural Beauty and chaired Governor Nelson Rockefeller's Conference on Natural Beauty in New York. His editorial help in rewriting the 1969 New York City Comprehensive Plan earned acclaim from the *New York Times* and the American Society of Planning Officials (Birch 1986).

The turbulent year of 1968 yielded a trio of environmental landmarks: Ian McHarg's *Design with Nature,* Garret Hardin's seminal article in *Science,* "The Tragedy of the Commons," and Whyte's *The Last Landscape.* Returning to the themes of his "Urban Sprawl" essay, *The Last Landscape* was Whyte's "bible" for the fast-spreading movement to save open space in metropolitan America. "Open space" was to conservationists of the 1960s what "anticongestion" was to early-twentieth-century progressives and what "sustainability" and "smart growth" are to environmentalists today. It embraced a variety of maladies from poorly planned development: loss of prime farmland, shortage of recreation space, urban flooding, pollution of surface water and groundwater, aesthetic blight, diminished sense of place, and isolation from nature. (Today we would add loss of biodiversity as well.) *The Last Landscape* offered a legal toolbox to combat urban sprawl, including cluster zoning, conservation easements, greenbelts, scenic roads, and tax abatements. Much of today's smart growth agenda was anticipated in *The Last Landscape* (which was republished in 2002 by the University of Pennsylvania Press).

If Whyte had confined himself to astute observation and witty commentary, his contribution would be notable but not lasting. What distinguished his legacy was his continuous agitation for practical improvement in urban design and land use, based on empirical observation and leading to measurable outcomes. For instance, Whyte helped reform the 1961 New York City's zoning provision that offered density bonus incentives to developers of new office or residential buildings in exchange for public amenities. With revisions suggested by Whyte, this approach to date has yielded more than five hundred privately owned and maintained public spaces, including street-level plazas, interior or covered public areas, arcades, and through-block gallerias. Planning lawyer Jerold S. Kayden (2000) has documented widespread problems with the accessibility and management of many of these spaces, yet in toto they make up an extraordinary legacy of shared spaces provided at private cost. (See Kayden's summary of his findings in his essay in this volume.)

Whyte's proudest accomplishment was the revitalization of Bryant Park in midtown Manhattan behind the New York City Public Library (Dillon 1996). By the late 1970s, the park had degenerated into a littered, seedy, and menacing space. Under Whyte's guidance as consultant and with funding from the Rockefeller Brothers Fund, the park was progressively restored, redesigned, replanted, and returned to its original use as a green oasis for the general public to enjoy. One of Bryant Park's most popular features is a plenitude of movable chairs, an idea bor-

Figure 2   A lively downtown plaza in Oakland, California, at lunch hour: a quintessential Holly Whyte urban scene. (Photo by R. H. Platt.)

rowed by Whyte from the Jardin du Luxembourg in Paris (see figure 2 in Eugenie Birch's essay in this volume). Ongoing management of the park today is entrusted to a "business improvement district" (BID), which levies a tax on surrounding real estate and holds special events to pay for enhanced maintenance. Such a novel public-private partnership between the city and the BID is consistent with Whyte's optimistic pragmatism.

From Park Forest in the 1950s to New York City in the 1980s, Whyte was a die-hard urban environmental determinist. He believed that the design of shared spaces greatly affects the interaction of people who encounter one another in those spaces and their resulting sense of well-being or discomfort in urban surroundings. Such interaction in turn helps shape the "success" of cities and suburbs as congenial or alien environments for the millions who inhabit them (figure 2). As the *New Yorker* architectural critic Paul Goldberger wrote in a foreword to a compendium of Whyte's writings edited by Albert LaFarge (2000, vii): "His objective research on the city, on open space, on the way people use it, was set within what I think I must call a moral context. Holly believed with deep passion that there was such a thing as quality of life, and the way we build cities, the way we make places, can have a profound effect on what lives are lived within those places."

U.S. Metropolitan Growth since the 1950s

Metropolitan America at the dawn of the twenty-first-century has sprawled far beyond the wildest imaginings of *The Exploding Metropolis* authors. Between 1950 and 2000, metropolitan statistical areas (MSAs) designated by the Bureau of the Census grew in number from 169 to 347, in population from 84 million to 226 million, and in size from 9 percent to about 18 percent of the land area of the conterminous United States (table 1).[1] "Suburbs" (areas within metropolitan areas other than central cities) grew from 55 million residents in 1950 to more than 141 million in 2000 and now are home to slightly more than one-half of the entire U.S. population. Metropolitan areas as a whole, including central cities, in 2000 accounted for four-fifths of the nation's population. By comparison, in 1960, central cities, suburbs, and nonmetropolitan areas each represented about one-third of the nation's population (figure 3).

Most metropolitan areas today are expanding spatially much more quickly than they are adding population. Between 1982 and 1997, the total extent of "urbanized areas," as delineated by the Bureau of the Census, increased by 47 percent while the nation's population grew by only 17 percent (Fulton et al. 2001). The Chicago area grew by 48 percent in population between 1950 and 1995 while its urbanized land area increased by 165 percent (Openlands Project 1998). Between 1970 and 1990,

Table 1 Changes in Metropolitan America, 1950–2000

	1950	*2000*
U.S. population	152 million	281 million
No. of metropolitan areas	169	347
Metropolitan population	84 million (55% of U.S.)	226 million (80% of U.S.)
No. of metropolitan areas > 1 million	14	39 (1990)
Population of metropolitan areas > 1 million	45 million (30% of U.S.)	125 million (50% of U.S.)
Metropolitan % of U.S. land area	9%	18%
Average metropolitan population density	407 persons/sq. mile	330 persons/sq. mile
Central city population	49 million (32% of U.S.)	85 million (30% of U.S.)
"Suburban" population*	35 million (23% of U.S.)	141 million (50% of U.S.)

* The Bureau of the Census does not use the term *suburb*. The term is colloquially used to represent all portions of metropolitan statistical areas (MSAs) outside of "central cities" (now called "principle cities").

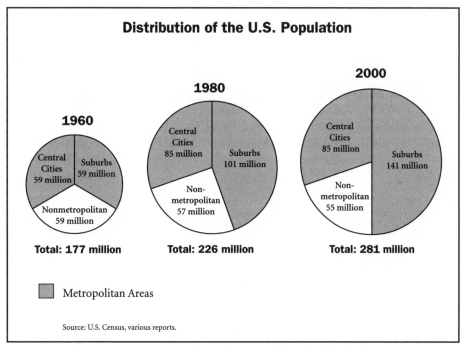

Figure 3 Distribution of the U.S. population among central cities, suburbs, and nonmetropolitan areas: 1960, 1980, 2000. (Source: University of Massachusetts Ecological Cities Project.)

the Los Angeles metropolitan population grew 45 percent while its urbanized land area expanded by 300 percent (table 2). Overall, the average density of urban America has declined from 407 persons per square mile in 1950 to 330 in 2000, but surprisingly, metropolitan expansion in the West is at higher average density (i.e., is less sprawling) than elsewhere in the United States, according to a Brookings Institution study (Fulton et al. 2001).

Between 1950 and 2000, "suburbs" tripled in population while central cities collectively gained only 73 percent. Even this comparison understates the actual shift away from older cities toward suburbs. The Bureau of the Census lists as "central cities" many new or greatly enlarged Sunbelt cities that are predominantly suburban in character, such as San Diego (75 percent population growth since 1970), Phoenix (145 percent), Los Angeles (27.9 percent), and Las Vegas (220 percent). These examples of "elastic cities" are defined by David Rusk (1999) as cities able to enlarge their geographic area through annexation of adjoining territory. The expansion in area and population of these elastic southwestern cities masks the heavy losses in the populations of many older northern cities whose boundaries are "inelastic." Between 1970 and 1990, Chicago lost about 17 percent of its total

Table 2   Expansion of Population and Urbanized Land Area in Four Metropolitan Areas, 1970–90

Metropolitan Area	Change in Population (%) 1970–90	Change in Urbanized Land (%) 1970–90
Chicago	+4	+46
Los Angeles	+45	+300
New York City	+8	+65
Seattle	+38	+87

Source: Porter 2000, fig. 2-3.

population, Minneapolis lost 19 percent, New York lost 6 percent, and Washington, D.C., shrank by nearly one-third.

Race and Poverty

When race is considered, the contrast is even starker. During the 1990s, the top one hundred cities in population experienced a 43 percent increase in Hispanic population (3.8 million people), a 6 percent increase in African American population (750,000 people), and a 38 percent rise in Asian population (1 million people). During the same decade, 2.3 million whites left those cities. In 1990, 52 percent of the combined populations of those one hundred cities was white; in 2000, that percentage had declined to 44 percent (Katz 2001). Journalist Ray Suarez in his book *The Old Neighborhood* summarizes experience in a few of the nation's largest cities as follows:

> Between 1950 and 1990, the population of New York stayed roughly level, the white population halved, and the black population doubled. As Chicago lost almost one million people from the overall count, it lost almost two million whites. As the population of Los Angeles almost doubled, the number of whites living there grew by fewer than ninety thousand. Baltimore went from a city of three times as many whites as blacks in 1950 to a city that will have twice as many blacks as whites in the year 2000. All this happened while the number of blacks in the United States has stayed a roughly constant percentage, between 11 and 13 percent. (Suarez 1999, 10)

Racial change is not necessarily bad if it results in greater access to decent housing and jobs for nonwhites. That, however, is not the case. To begin with, blacks are more likely to be poor than whites. In 1993, the percentage of white families below the federal poverty level was 9.4 percent, compared with 31.3 percent of black families. Both of these proportions had increased since 1979 when 6.9 percent of white families and 27.8 percent of black households were below the poverty level (U.S. Bureau of the Census 1995–96, table 752). Thus, with blacks making up a rising proportion of city population and poverty afflicting a rising proportion of black households and individuals, it follows that black poverty is heavily concentrated in central cities.

Yet this situation does not translate into improved housing or economic opportunities for lower-income nonwhites by virtue of living in cities. Housing in "ghetto" neighborhoods is notoriously dilapidated but nevertheless costly to rent because poor tenants seldom have anywhere else to turn. David Rusk (1999, 70–71) quotes a bitter indictment by Oliver Byrum, former planning director of Minneapolis: "Low-income people and poverty conditions are concentrated in inner city areas because that is where we want them to be. It is, in fact, our national belief, translated into metropolitan housing policy, that this is where they are supposed to be. Additionally, they are to have as little presence as possible elsewhere in the metropolitan area. . . . *Cheap shelter is to be mostly created by the devaluation of inner city neighborhoods*" (emphasis added).

Furthermore, poverty itself is not colorblind. According to Rusk (1999, 71), poor whites in metropolitan areas about equaled the total of poor blacks and Hispanics combined in 1990. Yet although three-quarters of the poor whites lived in "middle-class, mostly suburban neighborhoods," the same percentage of poor blacks and Hispanics inhabited inner-city, low-income neighborhoods.

Despite federal laws to protect civil rights, open housing, and equal opportunity, central cities are more racially and economically challenged than ever. Consider Hartford, the state capital of Connecticut, the wealthiest city in the United States after the Civil War and home to Mark Twain, Louisa May Alcott, Trinity College, and the Travelers Insurance Company. Hartford was recently described by the *New York Times* as "the most destitute 17 square miles in the nation's wealthiest state, and a city where 30 percent of its residents live in poverty. Only Brownsville, Texas, has a higher figure" (Zielbauer 2002).

Adding to the downward spiral of older central cities, new jobs have been predominantly created in suburban locations, thus requiring inner-city residents to have a personal vehicle for an often-lengthy reverse commute. In the case of Atlanta, the central city's share of the metropolitan job market dropped from 40 percent in 1980 to 19 percent in 1997. From 1990 to 1997, the central city gained only 4,503 new jobs, just 1.3 percent of all jobs created in the region during that period while 295,000 jobs, or 78 percent of all jobs, were added to Atlanta's northern suburbs (Bullard, Johnson, and Torres 2000, 10–11). Furthermore, poor public transportation may impede residents of low-income neighborhoods from even reaching jobs "downtown" or elsewhere within their own cities.

Edge Cities

Not only have jobs followed the white middle class to the suburbs, but much of the new economic activity outside the central city is likely to be concentrated in "edge cities" (Garreau 1991), also called "urban villages" (Leinberger and Lockwood

1986) or "mushburbs" (Platt 2004, 195). An edge city is a high-density complex of retail, office, hotel, entertainment, and high-end residential uses, typically situated near major interstate highway interchanges (e.g., the Burlington Mall area northwest of Boston), airports (the vicinity of Chicago's O'Hare International Airport) or rapid transit stations (Ballston on the Washington, D.C., Metro Orange Line). Like Holly Whyte's study of Park Forest in *The Organization Man,* Joel Garreau's *Edge City: Life on the New Frontier* explored the physical and human dimensions of this late-twentieth-century phenomenon. Using a definition of a newly developed cluster having at least five million square feet of office space and 600,000 square feet of retail space, among other criteria, Garreau identified more than two hundred edge cities in metropolitan areas across the United States. Astonishingly, he estimated that edge cities in 1991 contained two-thirds of all U.S. office space, thus eclipsing conventional urban "downtowns." The edge cities of New Jersey contained more office space than the financial district of Manhattan. South Coast Mall in Orange County, California, did more business in a day than did all downtown San Francisco (Garreau 1991, 5, 63).

How would Holly Whyte have felt about edge cities? He might have accepted them in certain respects—high-density land coverage, mingling of people in quasi-public spaces, casual eating facilities (the ubiquitous "Café Square"), and convenient pedestrian access (once the SUV is stowed in the parking structure)—but they have no "streets" or street life, which he revered. Commercial space is leased by formula set in shopping mall bibles, with little freedom for unorthodox retail uses or groups of similar businesses (e.g., a fortune teller next to a TGIF outlet or a row of fortune tellers). Rents are high, which excludes most low-volume or specialty stores, although space for cart vendors is sometimes allowed. The overriding characteristic of edge cities that would have probably vexed Whyte, however, is their "privateness." They represent the logical progression of the conventional shopping center where all space is managed directly or indirectly by the development company. It opens and closes at fixed times, it is usually clean and orderly, but free speech, unlicensed entertainment, odd behavior, and "undesirables" (Holly Whyte's pre-PC term) are subject to expulsion. The entire place is relatively "new," which means that it will all grow obsolete at the same time. Although buildings may be separately owned, the entirety is subject to an overriding master plan and there is no place for the unexpected. In short, *it is not a city.*

Nor is it even a town. Edge cities by definition are not governmental units and thus are private enclaves within larger units of local government. Involvement with that larger community, its schools and other civic life, may be very limited for edge city residents and employees. Although the edge city is likely a major source of property and sales tax revenue, it may be viewed by the local populace as an alien presence rather than as an integral part of "the community." Local authorities seldom turn down proposals for new edge cities or their smaller cousins, however.

Gated Communities

The trend toward privateness in U.S. metropolitan growth is nowhere more obvious than in the spread of "gated" residential communities. Gated communities are
subdivisions surrounded by literal and legal walls. Whereas streets, bike paths, and
recreation amenities in a traditional subdivision are conveyed to the local government and are open to the public, a gated subdivision retains control over these
features and restricts access to them. Gated communities often include golf courses,
tennis courts, and other membership amenities, funded out of homeowner assess-
ments and user fees, and open only to residents and their guests.

In their pioneering study of gated communities, planners Edward J. Blakely and
Mary Gail Snyder (1997, 7) estimated that there were by 1997 "as many as 20,000
gated communities, with more than 3 million units. They are increasing rapidly in
number, in all regions and price classes." They are most common in affluent outer
reaches of Sunbelt metropolitan areas, but large concentrations also are found in
wealthy suburbs of most larger cities. (See Blakely's essay in this volume.)

The gated community is "anti-Whyte." Its very gatedness and exclusion of the
nonapproved flies in the face of the proletarian democracy of the street celebrated
by Holly Whyte. Furthermore, the privateness of the home surroundings becomes
extended to privateness in all aspects of life: private school, private clubs, private
resorts, closely guarded places of work (in edge cities, perhaps), and a general dis-
trust of the outside world.

With jobs and homes broadly scattered across the metropolis, public transportation systems have often atrophied for lack of passengers and revenue, if they existed in the first place. The metropolitan workforce spends a growing percent of its
waking hours commuting (76 percent alone) on jammed freeways. In the Atlanta
area, the average driver travels thirty-four miles a day and spends sixty-eight hours
a year trapped in traffic gridlock, making Atlanta the fourth worst commuting
region in the United States behind Los Angeles, Washington, D.C., and Seattle
(Bullard, Johnson, and Torres 2000, 12). Since 1970, the nation's motor vehicles
have nearly doubled in number (not to mention size) while the population has
grown by 40 percent and road capacity increased by 6 percent (Seabrook 2002).
The more affluent commuters repeat the ordeal on weekends to reach the Hamp-
tons, Cape Cod, Maine, the Eastern Shore of Maryland, northern Michigan, the
Sierra foothills, and other supposed refuges from the "madding crowd."

Meanwhile, Whyte's beloved downtowns (except Manhattan and a few other
twenty-four-hour city centers) are conspicuously deserted at night. Although there
are many more plazas, mini-parks, and other social spaces, the daytime office
crowd vanishes to the suburbs and beyond on evenings and weekends. In many
cities, revival of downtowns has focused on attracting suburbanites and tourists
through megastructures such as sports stadiums, conference centers, casinos, and

"festival marketplaces" such as Boston's Quincy Market, Baltimore's Inner Harbor, and San Francisco's Embarcadero. Although these areas offer economic benefits, they often represent a reversion to urban gigantism in the spirit of urban renewal that was early decried by Whyte and Jacobs. Although the outdoor marketplaces (as in San Diego's Horton Plaza) may offer some sense of urbanity and spontaneity, their indoor elements in general are private downtown malls. Even in Manhattan, many public spaces lack users owing to design or management deficiencies (Kayden 2000). (See Kayden's essay in this volume.)

Suburbs have, of course, changed in many respects and today are unlikely to resemble the lily-white "organization man" suburbs of the 1950s (except that many are still white). Even Whyte's archetype Park Forest, Illinois, experienced deliberate racial change in the 1960s; today with its counterparts around the country it is a diverse community sociologically, if not economically (as documented by a recent film on Park Forest by James Gilmore, *Chronicle of an American Suburb*). Suburbs are also more diverse in terms of lifestyle and household status: the stereotypical nuclear family of television sitcoms in the 1950s like *Ozzie and Harriet* has been supplanted in many suburbs by increasing numbers of singles, elderly people, and gay and single-parent families. Both Ozzie and Harriet have jobs, if they are lucky, and may be divorced or separated with some sort of shared custody of the children. Suburban nonfamily households—mostly young singles and elderly living alone—outnumbered married couples with children according to the 2000 census (Frey and Berube 2002).

Another outdated stereotype, in more affluent communities at least, is the image of the vapid cultural life and humdrum retail and entertainment opportunities of suburbia. David Brooks in *Bobos in Paradise: The New Upper Class and How They Got There* (2000) satirizes the proliferation in suburbs of trendy coffee bars, health food outlets, multicultural galleries, and other services loosely associated with "urbanism." Yet although Starbucks and its ilk are solidly established in America's upscale suburbs and shopping malls, a mall is still a mall. Although they gain more ethnic flavor and cater to the wider diversity of suburbia itself, malls remain private enclaves where commercial occupancy and personal behavior are highly regulated and where the uniform building design, controlled climate, and background "elevator music" are similar from coast to coast.

When private malls are combined with gated residential compounds, private transportation, private schools, and private recreation clubs, there is little left in the contemporary metropolitan area that is "public" or "community-based." Robert D. Putnam's *Bowling Alone* (2000), which depicts the loss of "community" in America's cities and suburbs alike, stands in counterpoint to Whyte's Park Forest of the 1950s. Organization life at least offered a kind of togetherness, a temporary substitute for the traditional urban neighborhoods celebrated by Jane Jacobs and Ray Suarez, and the proverbial small towns of Norman Rockwell and the Archie comics. Yet even that is now diminished, as Brooks (2000, 238) observes: "Today few writers argue that Americans are *too* group oriented or *too* orderly. They are not complaining about Organization Man or the other-directed joiners. On the contrary, today most social critics are calling for *more* community, *more* civil society, *more* social cohesion."

Toward a More Humane Metropolis

All is not lost, however. Even as metropolitan America has become more populous, more sprawling, more exasperating, and more stratified, a subliminal countervailing trend is beginning to stir. In cities and suburbs across the United States, in both red states and blue, myriad local efforts are under way to make urban communities more amenable to people and nature, in short, to make them more "humane." This book and the conference that gave rise to it sample a few of these efforts as harbingers of the humane metropolis. This concept (a deliberate play on "The Exploding Metropolis") draws from and expands upon the work of William H. Whyte, in company with that of Jane Jacobs, Ian McHarg, Kevin Lynch, Ann Louise Strong, Charles E. Little, Tony Hiss, Ann Whiston Spirn, and many others. The phrase "humane metropolis" as used in this book means urban places that are more *green,* more *healthy and safe,* more *people friendly,* and more *equitable.*

Today, efforts to preserve and restore nature within urban regions are breaking new ground, so to speak. Some ecologists are finally beginning to specialize in urban ecology (Daily 1997), as in the long-term ecological research programs in Baltimore and Phoenix funded by the National Science Foundation (Collins et al. 2000; Grimm et al. 2000). American Forests, a nongovernmental organization based in Washington, D.C., is documenting the benefits of preserving tree canopy in the urban environment. The U.S. Forest Service and the Conservation Fund are promoting the concept of "green infrastructure." New forms and uses of city parks are documented in a thorough study by Peter Harnik for Trust for Public Land and the Urban Land Institute (Harnik 2000, summarized in his essay in this volume). Urban gardens are appearing in surprising places, such as on the roof of Chicago's City Hall. A joint project of Rutgers University and the Brooklyn Botanic Garden is testing ecological ways to cover landfills. (See the essay by Clemants and Handel, this volume.) Some suburban lawns are being relandscaped with native vegetation. Urban watersheds like the Mystic and Neponset in Boston, the Milwaukee River, and the San Diego River are being ecologically rehabilitated, at least in limited segments. Urban vacant lots are sprouting gardens in New Haven under the auspices of the Urban Resources Initiative, a joint venture of the Yale School of Forestry, the City of New Haven, and neighborhood organizations. The National Audubon Society is creating urban environmental education centers in Prospect Park in Brooklyn and Debs Park in Los Angeles in partnership with local public and private interests. A group named ARTScorpsLA under the guidance of artist Tricia Ward has created a beautiful neighborhood eco-park in Los Angeles called *La Coulebra,* turning discarded concrete rubble into art forms decorated by neighborhood children. Experience in European cities with urban regreening is described by planner Timothy Beatley in *Green Urbanism* (Beatley 1998, summarized in his essay in this volume). And so on!

The Ecological Cities Project (www.ecologicalcities.org) is a program of research, teaching, and outreach on the regreening and social revival of urban communities based at the University of Massachusetts Amherst. William H. Whyte had influenced my interest in cities and greenspaces since my early days as a staff attorney with the Openlands Project in Chicago. I invited him to be keynote speaker at a conference several colleagues and I held in Chicago in 1990, which led to the "prequel" to this book, *The Ecological City: Restoring and Preserving Urban Biodiversity* (Platt, Rowntree, and Muick 1994).[2] After Whyte's death in 1999, it was natural for our newly established Ecological Cities Project to celebrate Whyte's work and its echoes in contemporary urban places today. With nary a pompous syllable, he laid the foundation for what would later be termed growth management, sustainable development, smart growth, New Urbanism, and a host of other buzzword movements. We were successful in persuading the Lincoln Institute of Land Policy, one of the world's leading land use research and education centers, to support the project, with additional support provided by the Wyomissing Foundation, the National Park Service, the U.S. Forest Service, and Laurance S. Rockefeller.

On June 6 and 7, 2002, approximately three hundred urban design practitioners, writers, ecologists, grassroots activists, and students gathered in New York City for "The Humane Metropolis: People and Nature in the Twenty-First Century—A Symposium to Celebrate and Continue the Work of William H. Whyte." The conference was held at the New York University Law School in collaboration with several units at that university. Other New York cooperating organizations included the Municipal Art Society, the Project for Public Spaces, the Brooklyn Botanic Garden, and the Regional Plan Association. Keynote speakers included Amanda M. Burden, chair of the New York City Planning Commission; Adrian Benepe, New York City Parks Commissioner; and Carl Anthony of the Ford Foundation. Several of Holly's friends and associates, as well as his daughter, Alexandra Whyte, expressed personal tributes.[3] In addition, the University of Pennsylvania Press released a new edition of *The Organization Man* at the symposium.

The rest of the two-day conference explored a series of present and proposed initiatives around the country that perpetuate or expand on Holly Whyte's ideas on people, nature, and cities. Some of the initiatives were represented there and in this book. Urban greening and revitalization projects at various scales from inner-city gardens to regional parks and habitat restoration programs were presented by speakers from the New York area as well as Chicago, Milwaukee, Boston, Durham,

N.C., Portland, Oregon, and elsewhere. Other sessions examined such topics as ecological restoration, environmental education, and regreening the built environment. Although a number of topics such as green roofs, urban gardens, and brownfield remediation were not part of Whyte's own palette of topics, we assumed that if he were to rewrite *The Last Landscape* today, he would applaud and document such new issues and approaches.

The conference was a success according to comments received. Among our favorite was the following from Peter Harnik, director of the Trust for Public Lands' Green Cities Initiative (and contributor to this volume):

> I tend to be slightly on the critical, hard-to-please side of the analytical spectrum, so it's even more meaningful when I say that the "Humane Metropolis" was one of the best conferences I've ever attended.... The speakers were consistently terrific, the audience was wonderful, and the audio-visual was taken care of flawlessly. Most important, you are on the cutting edge of an up-and-coming topic that is given almost no attention by anyone else—since urban experts rarely talk about nature, and conservationists virtually never talk about cities.

We believe Holly Whyte would say "Amen"!

Notes

1. As of the 2000 census, MSAs other than in New England were designated as clusters of one or more counties anchored by a core consisting of one or more cities or an "urbanized area" containing at least fifty thousand inhabitants. Using counties as the building blocks of MSAs leads to inclusion of much nonurban land where counties are very large, as in Southern California. In New England, MSAs consisted of clusters of cities and towns instead of counties. Metropolitan area terminology and classifications have since been revised. (See Frey et al. 2004.)

2. Unfortunately, that book did not include a Whyte paper owing to his poor health at the time.

3. Holly Whyte's widow, the indomitable Jenny Bell Whyte, attended the entire symposium despite health issues; she passed away three months later on September 1, 2002.

References

Beatley, T. 1998. *Green urbanism: Learning from European cities.* Washington, DC: Island Press.

Birch, E. L. 1986. The observation man. *Planning,* March, 4–8.

Blakely, E. J., and M. G. Snyder. 1997. *Fortress America: Gated communities in the United States.* Cambridge, MA: Lincoln Institute of Land Policy and Brookings Institution Press.

Brooks, D. 2000. *Bobos in paradise: The new upper class and how they got there.* New York: Simon and Schuster.

Bullard, R. D., G. S. Johnson, and A. O. Torres, eds. 2000. *Sprawl city: Race, politics, and planning in Atlanta.* Washington, DC: Island Press.

Burgess, E. W. 1925. The growth of the city: An introduction to a research project. In *The city,* ed. R. E. Park, E. W. Burgess, and R. D. McKenzie. Republished, Chicago: University of Chicago Press, 1972.

Chen, D. D. 2000. The science of smart growth. *Scientific American* 283(6): 84–91.

Collins, J. P., et al. 2000. A new urban ecology. *American Scientist* 88(5): 416–25.

Daily, G. C., ed. 1997. *Nature's services: Societal dependence on natural systems.* Washington, DC: Island Press.

Diamond, H. L., and P. F. Noonan. 1996. *Land use in America.* Washington, DC: Island Press.

Dillon, D. 1996. The sage of the city, or how a keen observer solves the mysteries of our streets. *Preservation,* September–October, 71–75.

Editors of *Fortune.* 1957. *The exploding metropolis.* Garden City, NY: Doubleday Anchor Books. Republished, Philadelphia: University of Pennsylvania Press, 2002.

Frey, W. H., and A. Berube. 2000. City families and suburban singles: An emerging household story from Census 2000 (Census 2000 Series Monograph). Washington, DC: Brookings Institution.

Frey, W. H., J. H. Wilson, A. Berube, and A. Singer. 2004. *Tracking metropolitan America into the 21st century: A field guide to the new metropolitan and micropolitan definitions.* Washington, DC: Brookings Institution.

Fulton, W., R. Pendall, M. Nguyen, and A. Harrison. 2001. Who sprawls most? How growth patterns differ across the United States (Survey Series Monograph). Washington, DC: Brookings Institution.

Garreau, J. 1991. *Edge city: Life on the new frontier.* New York: Doubleday Anchor.

Gill, B. 1999. The sky line: Holding the center. *New Yorker* 65(3): 99–104.

Glazer, N. 1999. The man who loved cities. *Wilson Quarterly* 23(2): 27–34.

Gottmann, J. 1961. *Megalopolis: The urbanized northeastern seaboard of the United States.* Cambridge, MA: MIT Press.

Grimm, N. B., et al. 2000. Integrated approaches to long-term studies of urban ecological systems. *Bioscience* 50(7): 571–84.

Hardin, G. 1968. The tragedy of the commons. *Science* 162: 1243–48.

Harnik, P. 2000. *Inside city parks.* Washington, DC: Urban Land Institute and Trust for Public Land.

Jackson, K. T. 1989. *Crabgrass frontier: The suburbanization of the United States.* New York: Oxford University Press.

Jacobs, J. 1957. Downtown is for people. In *The exploding metropolis,* ed. Editors of *Fortune.* Garden City, NY: Doubleday Anchor.

———. 1961. *The death and life of great American cities.* New York: Random House Vintage.

Katz, B. 2001. The new metropolitan agenda. Speech to the Smart Growth Conference, June 9.

Kayden, J. 2000. *Privately owned public space: The New York City experience.* New York: Wiley.

LaFarge, A., ed. 2000. *The essential William H. Whyte.* New York: Fordham University Press.

Leinberger, C. B., and C. Lockwood. 1986. How business is reshaping America. *Atlantic Monthly* 258(4): 43–52.

Little, C. E. 1963. *Challenge of the land.* New York: Open Space Institute.

McHarg, I. 1968. *Design with nature.* New York: Garden City Press.

Openlands Project. 1998. *Losing ground: Land consumption in the Chicago region, 1900–1998.* Chicago: Openlands Project.

Platt, R. H. 2000. Ecology and land development: Past approaches and new directions. In *The practice of sustainable development,* ed. D. R. Porter. Washington, DC: Urban Land Institute.

———. 2004. *Land use and society: Geography, law, and public policy.* Rev. ed. Washington, DC: Island Press.

Platt, R. H., R. A Rowntree, and P. C. Muick, eds. 1994. *The ecological city: Preserving and restoring urban biodiversity.* Amherst: University of Massachusetts Press.

Porter, D. R., ed. 2000. *The practice of sustainable development.* Washington, DC: Urban Land Institute.

Putnam, R. D. 2000. *Bowling alone: The collapse and revival of American community.* New York: Simon and Schuster.

Riis, J. A. 1890. *How the other half lives.* Republished, Williamstown, MA: Corner House Publishers, 1972.

Rome, A. 2001. *The bulldozer in the countryside: Suburban sprawl and the rise of American environmentalism.* New York: Cambridge University Press.

Rusk, D. 1999. *Inside game, outside game: Winning strategies for saving urban America.* Washington, DC: Brookings Institution Press.

Seabrook, J. 2002. The slow lane: Can anyone solve the problem of traffic. *New Yorker,* 2 September, 120–29.

Suarez, R. 1999. *The old neighborhood: What we lost in the great suburban migration: 1966–1999.* New York: Free Press.

Teaford, J. C. 1993. *The twentieth-century American city,* 2nd ed. Baltimore: Johns Hopkins University Press.

U.S. Bureau of the Census. 1995–96. *American Almanac: Statistical Abstract of the United States.* Austin, TX: The Reference Press.

Whyte, W. H., Jr. 1956. *The organization man.* New York: Simon and Schuster. Republished Philadelphia: University of Pennsylvania Press, 2002.

———. 1957. Urban sprawl. In *The exploding metropolis,* ed. Editors of *Fortune.* New York: Doubleday.

———. 1959. *Securing open space for urban America: Conservation easements.* Washington, DC: Urban Land Institute.

———. 1962. *Open space action: A report to the Outdoor Recreation Resources Review Commission.* Washington, DC: U.S. Government Printing Office (ORRRC Study Report 15).

Whyte, W. H. 1968. *The last landscape.* New York: Doubleday. Republished, Philadelphia: University of Pennsylvania Press, 2002.

———. 1980. *The social life of small urban spaces.* Washington, DC: The Conservation Foundation.

———. 1988. *City: Rediscovering the center.* New York: Doubleday.

Winks, R. W. 1997. *Laurance S. Rockefeller: Catalyst for conservation.* Washington, DC: Island Press.

Zielbauer, P. 2002. Poverty in a land of plenty: Can Hartford ever recover? *New York Times,* 26 August, 1, B4.

PART ONE

"The Man Who Loved Cities"

Among many tributes paid to William H. "Holly" Whyte after his death in 1999, Norman Glazer (1999) characterized him in the *Wilson Quarterly* as "the man who loved cities . . . one of America's most influential observers of the city and the space around it." It is fitting to devote Part I of this book to personal recollections of this perceptive urbanist written by several people who knew him well in different capacities and at different periods in his career. Ann Louise Strong, emerita professor of law at the University of Pennsylvania, worked closely with Whyte on the pathbreaking open space plan for the Brandywine Valley in the mid-1960s. Another Penn colleague, the late landscape architect Ian McHarg, drew strongly on the Brandywine plan in his seminal 1968 book, *Design with Nature.* In his turn, Whyte touted McHarg's ideas in his own 1968 classic, *The Last Landscape.*

Also in the mid-1960s, Charles E. Little had the good fortune to encounter Whyte as a board member of the New York Open Space Institute, of which Little was chief executive officer. Little eschewed being an "organization man" after reading Whyte's 1956 book of that title and instead applied his talents to making metropolitan New York a more habitable and "humane" place in which to live and work.

At the Open Space Institute, Little authored two books on land conservation: *Stewardship* and *Challenge of the Land* (which in turn influenced this editor's first effort, *Open Space in Urban Illinois,* Platt 1971). Subsequently, in the Washington, D.C., area and now in New Mexico, Little has contributed his literary and practical experience to many land-saving and regional planning efforts such as farmland preservation, "greenline parks," and sacred Native American sites.

Eugenie L. Birch, while a Hunter College urban planning professor and member of the New York City Planning Commission, collaborated with Whyte in his "Street Life Project"—the basis for his book and film, *The Social Life of Small Urban Spaces,* and his capstone book, *City: Rediscovering the Center*—during the 1980s. Birch would build on Whyte's passion for lively city centers in her own research on downtown revitalization.

Thomas Balsley, a practitioner of people-oriented urban design, has had many opportunities to apply lessons taught by Whyte. As a personal friend and sometime collaborator, Balsley helped realize Whyte's visions as to what works or fails in terms of people interaction and enjoyment in shared urban spaces.

Albert LaFarge became a Holly Whyte fan toward the end of Whyte's career. As a frequent visitor to the East Ninety-fourth Street brownstone where the

Whytes lived, LaFarge assumed the role of Boswell to Holly's Johnson. The result was *The Essential William H. Whyte* (LaFarge 2000), which draws from all Whyte's writings the very best of his wit and wisdom.

References

Glazer, N. 1999. The man who loved cities. *Wilson Quarterly* 23(2): 27–34.

LaFarge, A., ed. 2000. *The essential William H. Whyte.* New York: Fordham University Press.

Little, C. E. 1961. *Stewardship.* New York: Open Space Action Committee.

———. 1963. *Challenge of the land.* New York: Open Space Institute

McHarg, I. 1968. *Design with nature.* New York: Garden City Press.

Platt, R. H. 1971. *Open space for urban Illinois.* DeKalb: Northern Illinois University Press.

Whyte, W. H. 1968. *The last landscape.* Garden City, NY: Doubleday. Republished, Philadelphia: University of Pennsylvania Press, 2002.

———. 1980. *The social life of small urban spaces.* New York: Project for Public Spaces.

———. 1988. *City: Rediscovering the center.* Garden City, NY: Doubleday.

Whyte on Whyte

A Walk in the City

Eugenie L. Birch

William H. Whyte *(The Observation Man)* left a remarkable body of writing that addressed three principal aspects of the United States after World War II:

1. The sociology of large organizations and their new suburban habitats (*The Organization Man*, 1956)
2. Suburban land use and sprawl (two essays in Editors of *Fortune; The Exploding Metropolis*, 1957; *Securing Open Space for Urban America: Conservation Easements*, 1959; *Cluster Development*, 1964; and *The Last Landscape*, 1968)
3. The functions and design of public spaces in urban settings (*The Social Life of Small Urban Places*, 1980; *City: Rediscovering the Center*, 1988).

Only today, as we are rebuilding lower Manhattan and other downtowns while confronting runaway suburban sprawl across the nation, are we realizing the prescience of this remarkable urbanist and his work. His understanding that economic concentration and population density at the center of a region is the key to conserving land at its periphery made him a pioneer of today's "smart growth" movement. Furthermore, he provided the theory and techniques for achieving model land use arrangements that contemporary city planners and metropolitan policy makers now vigorously promote.

A keen and sensitive observer of his surroundings, Whyte first approached an issue intuitively, but once he had a handle on it, he pursued it in depth, forging his own research methods. He read widely on the given subject, he talked to experts, but most important, he did field research, always questioning the so-called conventional wisdom. When he finally synthesized it all, he provided, in every instance, a new take on the selected topic that blended intelligence, wit, and common sense. This process was the source of his originality because, like Frederic Law Olmsted and other "enlightened amateurs," he was not trained in the field that he would help transform, namely urban planning and design. He picked up the basics as he went along: markets and economics from his experience as an editor at *Fortune* magazine; sociology from studying corporate life in Park Forest, Illinois; and land use planning from observing suburban development around his childhood home in Chester County, Pennsylvania, and from his consultancy on New York City's comprehensive plan.

Figure 1   William H. Whyte in later life in his beloved midtown Manhattan. (Photo by permission of Enrico Ferorelli.)

Whyte was a modern Renaissance man, a practical humanist, who became an authority on the burning issues of his day through self-education and intelligent observation (figure 1). In addition, he took his knowledge beyond articles, chapters, and books; he translated it into legislation and principles of urban design practice. He trained a generation of influential scholars and civic leaders. His insights on the use and design of urban public space are still fresh today. A year after his death in 1999, Fordham University Press printed a compendium of some of his best writings in *The Essential William H. Whyte* (LaFarge 2000), and the University of Pennsylvania Press reprinted *The Last Landscape* in 2001 and *The Organization Man* in 2002 (it also plans to reissue *The City: Rediscovering the Center*). *The Social Life of Small Urban Spaces* (1980) is marketed by the Project for Public Spaces, a design firm that Whyte helped found. (See Andrew G. Wiley-Schwartz's essay in this volume.)

In 1985, Whyte was elected an honorary member of the American Institute of Certified Planners (AICP) for his "outstanding contribution . . . to the development of the planning profession." (Also honored at the time were Lewis Mumford, the distinguished urban historian, and James Rouse, the builder of Columbia, Maryland, and originator of the concept of "festival malls.") The AICP cited

Whyte's "constructive influence on understanding subdivision growth, conversion of open space, cluster development, urban beautification, revitalization of central cities and the social life of small urban places" (Singer 1985).

At this time I was invited to write the article (Birch 1986) about Whyte for *Planning* magazine that later appeared under the title "The Observation Man" (a title the *New York Times Magazine* borrowed in a short profile of Whyte in its "People of the Millennium" issue of January 2, 2000). On a crisp, clear autumn day in 1985, Holly Whyte invited me to his office high above midtown Manhattan in Rockefeller Center where at age sixty-nine he was actively consulting and writing under the sponsorship of his longtime friend Laurance S. Rockefeller. We spent a few intense hours discussing his life, ideas, and many projects while overlooking Whyte's world: the glittering buildings of midtown Manhattan, the shimmering Hudson River, the New Jersey waterfront, and the hazy hills of the Garden State beyond. He had spent a lifetime puzzling over the various elements that made up the regional landscape, and he was eager to share his accumulated insights.

Whyte recalled his postgraduate days, portraying a raw Princeton University English major turned traveling salesman, peddling Vicks VapoRub during the Depression. He admitted that World War II had rescued him from that life. As a Marine intelligence officer, he began to develop a lifelong interest in geographic data, as later recorded in his final memoir, *A Time of War: Remembering Guadalcanal, A Battle without Maps* (Whyte 2000). After the war, he secured an editorial job at *Fortune* magazine, which at that time allowed a very broad interpretation of business journalism. He relished the extended time spent in the new white-collar Chicago suburb of Park Forest, Illinois, researching the series on the modern corporate worker that would be a key element of *The Organization Man,* his most successful book.

In the early 1950s, Whyte and several colleagues, including Jane Jacobs, wrote a series of articles that were republished as *The Exploding Metropolis* (Editors of *Fortune* 1957). That small book would become required reading for many planning students; Charles Abrams, head of the Columbia University planning program, considered it the best work in the field.

In the next few years, Whyte's urban philosophy would broaden and mature. In the early 1960s, his role as consultant to the Outdoor Recreation Resources Review Commission, chaired by Laurence S. Rockefeller, the philanthropic conservationist, stimulated his interest in techniques to retain open space in the process of suburban development, resulting in his professional reports on "conservation easements" and "cluster development."

In the mid-1960s, Whyte was retained by Donald H. Elliott, chair of the New York City Planning Commission, to overhaul editorially the city's draft Comprehensive Plan. Finding that the plan contained masses of data but lacked a clear message, Whyte rewrote much of it, producing a document that the *New York*

Times described as "probably the most clearly written plan ever published." The plan's "Critical Issues" section, which best reflects Whyte's graceful writing style and enthusiasm for urban life, begins:

> There is a great deal that is very right with New York City. As never before it is the national center of the United States . . . there is more of everything here that makes a city jump and hum with life—more of different kinds of people, more specialized services, more stores, more galleries, more restaurants, more possibilities of the unexpected. Here is the engine. And it is getting stronger. . . . Concentration is the genius of the City, its reason for being, the source of its vitality and its excitement. We believe the center should be strengthened, not weakened, and we are not afraid of the bogey of high density. (New York City Planning Commission 1969, vol. 1, p. 5)

In helping rewrite the 1969 Comprehensive Plan, Whyte encountered "incentive zoning," a technique added to the New York City zoning ordinance in 1961 that gave developers extra floor space in exchange for providing an urban plaza or public arcade at their expense. Employed extensively along the rapidly developing Sixth Avenue, the results, according to Whyte (and many other critics) were mixed. Although the city had gained additional open space, the sites were, in general, disappointing. They were poorly designed, unattractive, and, as a consequence, underused. The discouraged planners, Whyte related, were ready to eliminate the incentives, but he cautioned them "not to throw the baby out with the bathwater."

Convinced that the law could stand if improved, he set out to discover what was needed. Following his by now-proven research method of information gathering, observation, and synthesis, he established the Street Life Project, based at Hunter College and funded by the National Geographic Society, the Rockefeller Brothers Fund, and others. His eyes lit up as he described his techniques: the time-lapse photography, the miles of film footage to review, and the joy of finding behavioral patterns. He clearly loved "spying" on his urbanites. He was amused by their spontaneous street conversations and the variety of their interchanges on street corners and at building entrances.

Characteristically, he used the word *schmoozing* to typify street conversations. A stiff scholar would never have used such a term, but Whyte was not stiff. He was, however, systematic: he gathered empirical data and translated the information into an organized set of planning principles that New York City would incorporate into its 1975 zoning ordinance revision. Other cities would follow suit. This work was not guesswork; his greatest strength was his extreme attention to detail—exact measurements of the width of a sitting ledge, the amount of sitting space measured in linear feet related to the square footage of a plaza—to arrive at appropriate legislative formulas.

As of 1985, ten years after the zoning revisions, he felt strongly about the need to evaluate and refine these ideas. No one would take him up on his idea until the mid-1990s, when the New York Department of City Planning collaborated with

Harvard professor Jerold S. Kayden and the Municipal Art Society to inventory and assess the entire stock of incentive-based public space. (See Kayden's essay in this volume.)

What Holly Whyte wanted most was to craft the outdoor elements of downtown so that they could support and enhance the processes that make "the city jump and hum with life." His larger purpose was to create an environment that would support urban density, the engine powering the center and sustaining the surrounding region (figure 2). To a newspaper writer, he articulated his aims: "What makes the jostling, bustling, elbow-to-elbow belly dance of life in Manhattan bearable, are small amenities like open spaces with movable chairs and food kiosks, sidewalks wide enough to accommodate crowds, stairs that are easy to climb" (Croke 1989).

Having completed his basic open space analysis, Whyte now embarked on other refinements and causes. For example, he was concerned about new high-rise construction and its effect on preexisting urban plazas. Pouring over sunlight and shadow studies for a particularly offensive building, he succeeded in convincing the city government to reduce its height despite its already being under construction. Its shadow would have wrecked havoc with a nearby plaza. (See the essay by Mary V. Rickel Pelletier in this volume.) In addition, for all his progrowth talk, Whyte also appreciated how the mix and texture of buildings of different age and style enriched the urban environment, and thus he helped found the New York City Landmarks Conservancy. Finally, he was at that moment trying to figure out how to convince the federal Internal Revenue Service that scenic easements in urban areas should have favorable tax treatment.

In our 1985 interview, Whyte was impressive for his mental agility, practicality, and humor. He seemed to love unraveling knotty problems. He clearly enjoyed the complexities of urban life, especially the interplay between regulation, development, and human behavior. Reflecting his consulting excursions to cities very different from New York—such as Detroit, Dallas, Minneapolis, and Tokyo—he was sensitive to geography and climate, size of city, and internal location patterns. In this last phase of his career, Whyte was a tireless advocate of healthy, busy downtowns wherever they might be located.

To provide some photographs for the *Planning* article, Whyte suggested taking a tour of midtown Manhattan so that he could demonstrate why some urban places were successful and others not. On the appointed day, the photographer and I were to meet him at Rockefeller Center and from there we would visit half a dozen places. The day dawned bright and bitterly cold—the temperature was well below zero and the wind was whistling—but Whyte was undaunted. "The weather is perfect," he said. "Let's go."

The first stop, Paley Park in midtown, was deserted but still very beautiful. There he noted how the entry steps would draw people in and pointed out the composition of the trees, food concession, and movable chairs and how the waterfall

Figure 2 (*Top*) Relaxing on Holly Whyte's movable chairs in New York's Bryant Park. (*Bottom*) Socializing in the sun, Bryant Park. (Photos by R. H. Platt.)

muffled the street noise. Walking farther, he demonstrated the correct ledge width by perching on the edge of a window frame and explained that it would be lined with sitting people on a nice sunny day. Then we moved on to the IBM Plaza, a large indoor public space that was well populated that day, although someone had removed some of the chairs, to Whyte's dismay. At Phillip Morris Plaza, another interior space, elements embodied perfection in Whyte's estimation: the Whitney Museum had lent a whimsical sculpture, and tables and chairs filled the area. While sipping a cappuccino, he playfully conversed with one of the dancing statues.

The best part of the whole morning had occurred a little earlier. While walking down Madison Avenue, Whyte pointed out a man who was walking rapidly down the west side of the street. "Just watch," Whyte whispered. "He's going to jaywalk to the other side on a diagonal." Well, within seconds, that is just what the man did. Whyte knew his city and its habits.

References

Birch, E. L. 1986. The observation man. *Planning,* March, 4–8.

Croke, K. 1989. The city in black and Whyte, author stands on corner watching world go by. *New York Daily News,* 15 February, 36.

Editors of *Fortune.* 1957. *The exploding metropolis.* New York: Doubleday.

LaFarge, A., ed. *The essential William H. Whyte.* New York: Fordham University Press.

New York City Planning Commission. 1969. *Comprehensive plan for New York City.* New York: New York City Planning Commission.

Singer, S. L. 1985. William H. Whyte named honorary member of the American Institute of Certified Planners. Press release. Washington, DC: American Planning Association.

Whyte, W. H. 1956. *The organization man.* New York: Simon and Schuster. Republished 2002 with a Foreword by Joseph Nocera. Philadelphia: University of Pennsylvania Press.

———. 1959. *Securing open space for urban America: Conservation easements.* Technical Bulletin 36. Washington, DC: Urban Land Institute.

———. 1964. *Cluster development.* New York: American Conservation Association.

———. 1968. *The last landscape.* New York: Doubleday. Republished 2001 with a Foreword by Tony Hiss. Philadelphia: University of Pennsylvania Press.

———. 1980. *The social life of small urban places.* New York: Project for Public Spaces.

———. 1988. *City: Rediscovering the Center.* New York: Doubleday.

———. 2000. *A time of war: Remembering Guadalcanal, a battle without maps.* New York: Fordham University Press.

Holly Whyte's Journalism of Place

Charles E. Little

"So let us be on with it. . . . If there ever was a time to press for precipitate, hasty, premature action, this is it." These words are from the penultimate paragraph of *The Last Landscape,* Holly Whyte's roundup of how, and why, we ought to preserve metropolitan open space. Not later, but now. Not after great long studies, but now.

One day in the deep, dark 1960s, Stanley Tankel, the estimable chief planner at the Regional Plan Association in New York, invited Holly Whyte and me to lunch at the Harvard Club. Holly was writing his landscape book at the time and expostulating about "action," which was his favorite word. My role in this conversation, as the young executive director of the Open Space Action Committee, of which Stanley and Holly were board members, was to shut up and listen.

"Well," says Stanley, who was feeling grouchy, "you know what planners think about *that.*"

"Okay, what?"

"We say: *Action drives out planning.*"

"Exactly," says Holly, a grin splitting his great long face.

In those days, open space preservation was a very big deal. It was the means by which "the civics," as Stanley called local activists, could mitigate the headlong rush to develop or pave over or redevelop (taller, uglier) every square inch of metropolitan land. The race for open space (including the inner-city space opened up courtesy of dynamite, wrecking balls, bulldozers, and cranes) was on, but there wasn't enough money in the world for conservationists to purchase and set aside threatened lands in behalf of nature, human and otherwise.

It was Holly's great contribution to get the civics to understand that the lack of money was immutable, and to give them the tools to save the land anyway. Today, open space preservation has a full kit of screwdrivers, levers, and wrenches, virtually all of them—cluster development, easements, land philanthropy, tax strategies, greenways, transfer of development rights, and a whole lot more—in use because of Holly Whyte.

Yet his was the work not of a professor of geography, but of a journalist: a *Fortune* magazine editor when the writers for the magazines of Time, Inc., of which *Fortune* was the classiest, set the standards for everyone else. And he was the author of *The Organization Man,* a best seller that had a profound influence, and still does, on people of a certain age, including me.

In the 1960s when I became the executive director of the Open Space Action

Committee (OSAC) (it was Holly who supplied not only the name of the organization but the intellectual foundation for our work), I was a refugee from Madison Avenue. I had retired from an advertising agency at the advanced age of thirty-two in substantial part because I had read *The Organization Man* and had concluded that I did not want to be one when I grew up. So by the time I arrived at OSAC in 1964, I was overjoyed to find that William H. Whyte was on the board. One time, over drinks at a bar somewhere, I told Holly that he had changed my life.

He turned his long-suffering face toward me. "Don't ever say that to me again, Little," he said. "I am not going to be responsible for whatever dumb choices you make." So we had another beer. I just kept my counsel and decided to adapt (steal?) Holly's open space–saving ideas for my own work, starting with a book called *Stewardship*, which was instrumental in saving thousands of acres of open space in the New York metropolitan region. (The story of that program is told in *The Last Landscape*.) Then I used his ideas in a book called *Challenge of the Land*, which stayed in print for seventeen years through several editions. More recently, I ripped him off again in my 1989 book *Greenways for America*, finding that in fact Holly was the first to popularize this idea.

The point is not that Holly invented all the land-saving gadgets of which he wrote, but that he knew how to contextualize them, how to furnish the handles so that nonspecialist readers would understand their importance. This brilliant foray into open space journalism began with conservation easements, an otherwise dry and recondite topic that Holly presented in, of all places, *Life* magazine, in those days (1959) the premier popular magazine when magazines were the at the top of the mass media heap.

Then came the dynamite government report in 1962. Nothing like it had been seen before; it was number 17 in the Outdoor Recreation Resources Review Commission study series entitled Open Space Action, which read like, well, *Life*. (The commission was set up by President Dwight D. Eisenhower in 1958 and led by Laurance S. Rockefeller.) Then came *Cluster Development*. And then came the whole ball of open space wax in *The Last Landscape*. Without his skills as a journalist, I doubt that the techniques Whyte proposed in these and subsequent writings would have had anywhere near the effect, for they were aimed at a nonspecialist audience, over the shoulder as it were of those who make and influence land use decisions. When an idea is presented to 6,800,000 *Life* readers—or even lesser amounts in trade books and important government reports—the message sent cannot be ignored.

So today, the question is, Where are the new Hollys? Where are the land conservation writers to whom attention must be paid? We can name a few, but are they as influential as William H. Whyte? And if not, why not? Surely there ought to be a whole lot more who can carry on in the high-powered tradition of Holly Whyte, Jane Jacobs, Lewis Mumford—journalists all.

I am not talking about nature writing here: there's plenty of that, maybe too much. What Holly knew was reporting, and the importance of people and their stories so as to get ideas across and encourage others to take action.

Yet this kind of writing is not just of matter of interviewing or even rhetorical skill, as important as they may be. It also has to do with vision. Maybe I am getting old and cranky or have lived too long in the wrong place (not far from Albuquerque, whose leaders, almost to a person, would like it to become Los Angeles, smog and all). One thing that Holly had, and inspired in others, however, was a sense of democratic possibility in making the good place. I have a theory that the vision of the good place that came out of 1930s progressivism, the New Deal, and the Works Progress Administration cultural programs helped give the men in foxholes and on the beaches, like Holly, a reason not only to survive, but to prevail, and to come home, and to do good work.

One time, Holly made a notation in a manuscript on the egregious loss of metropolitan open space that I had sent him for review. "It is not necessary to be cynical," he wrote. I have never forgotten that and can picture the note in my mind even now.

I would submit that what's lacking today in the journalism of open space and urban place is vision, the visionary sense of progressive possibility. For the most part, the response to dehumanized metropolitan areas and urban cores these days is limited either to despairing jeremiads or to arcane discussions of the systems to curb the excesses of developers and intellectually challenged city officials who support them. The jeremiads don't work, and the corrective nostrums offered by planners may sound realistic—infill, adaptive reuse, intermodal transportation—but their expression, most often, is pinched and unimaginative. It fails to inspire. Surely we can do better than that. Holly's mind was infused with pictures of a humane city and a beautiful countryside. So let us not overlook the need for visioning the good place and then acting hastily, precipitately, and prematurely to make it happen.

In the end, we cannot succeed without vision, which was Holly's great gift as a writer and as a conservationist. For without vision, said Isaiah, another good writer, the people perish.

The Energizer

Ann Louise Strong

My first acquaintance with Holly Whyte goes back to the early and mid-1960s. At that time, he was overseeing and editing the multivolume Outdoor Recreation Resources Review Commission report. I was writing a book for the Urban Renewal Administration (URA), *Open Space for Urban America* (Strong 1961), to publicize and promote the URA's newly enacted and funded program for preservation of urban open space. It was the time when open space arose to importance on the national agenda, with Lady Bird Johnson our cheerleader in the White House. Holly, then as always, was an articulate, informed, and vigorous proponent who energized a groundswell of enthusiasm.

Holly and I became good friends and compatriots in the battle to publicize the availability of tools short of fee simple acquisition in the growing struggle to manage sprawl. These tools could preserve open space in private hands and private use at a cost far below that of public purchase. Holly was promoting conservation easements, speaking vigorously and often to conservation groups across the United States. I was working at the University of Pennsylvania Law School, with Jan Krasnowiecki on development of an alternative approach: compensable regulations (Strong and Krasnowiecki 1963). Thanks to the Ford Foundation, and to Gordon Harrison in particular, we enjoyed the advice of Holly as our consultant.

Holly was a native of West Chester, Pennsylvania, and a committed advocate of efforts to preserve the rich farmland and scenic setting of Chester County's Brandywine River valley. My family and I were newcomers to Chester County, settling there only in 1959, but I soon became equally dedicated to the task of protecting the Brandywine's exceptional resources. I gathered a group of nationally renowned resource scientists, planners, and economists to develop a plan for the protection of the water resources of the Upper East Branch of Brandywine Creek through management of land use. We were committed to reliance on less than fee controls for the plan's implementation. We gained financial and policy support from, among others, the Ford Foundation, including the appointment of Holly as one of our consultants. We spent several years of technical study, while involving local leaders in the evolving plan. Holly was often amongst us, speaking at meetings and offering advice. My admiration of his keen mind and acute sense of public sentiments grew and grew. Although the plan (Strong et al. 1968; Strong 1971) did not receive sufficient municipal support to be carried out, it did serve as a model for many subsequent efforts in the Brandywine and elsewhere. For instance, the

Brandywine Conservancy now holds conservation easements on more than 35,000 acres and is a major force in the Philadelphia region for protection of urban open space.

Holly and I did not work together again, but we continued to see each often, many times when speaking at conferences. Such was our final, sad get-together. Holly was to be the keynote speaker in Chicago at Rutherford Platt's "Symposium on Sustainable Cities" in 1990. I also was on the program, and the morning of our presentations we enjoyed breakfast together. Holly had a bad cold but otherwise was, as ever, full of tales of achievements in preservation from around the country, many of which he had fostered. Then, shortly after the conference, he suffered a debilitating attack that marked the end of his wonderful, inspiring participation in a world that many of us shared.

References

Strong, A. L. 1961. *Open space for urban America.* Washington, DC: Urban Renewal Administration.

———. 1971. *Private property and the public interest: The Brandywine experience.* Baltimore: Johns Hopkins University Press.

Strong, A. L., R. E. Coughlin, J. C. Keene, L. B. Leopold, B. H. Stevens, et al. 1968. *The plan and program for the Brandywine.* Philadelphia: Institute for Environmental Studies, University of Pennsylvania.

Strong, A. L., and J. E. Krasnowiecki. 1963. Compensable regulations for open space: A means of controlling urban growth. *Journal of the American Institute of Planners* (May).

Sowing the Seeds

Thomas Balsley

Holly Whyte's reach and influence were as diverse and unpredictable as the silent constituency he observed and championed. Some listened and were immediately persuaded; others nodded their heads approvingly but continued with their preconditioned behavior (only to be slowly converted after many observations and, in some cases, failures); and many others became disciples, joining the immediate family and sowing the seeds with actual practice.

My relationship with Holly fell into this last category, based mostly on my personal need to act, not talk. In many respects, Holly's simple, straightforward, and commonsense observations were the perfect formula and approach for me, and others like me, who were subconsciously searching for a counterbalance to the esoteric and theoretical preaching du jour. We could get our arms around these simple time-tested and approachable principles, as could our clients. Most important, they were conveyed to us in friendly constructive language—without judgment—in a structure that could be used in our collaborative pursuit of a better urban condition through the designed environment.

I can easily cite those facets that attracted me to landscape architecture: natural systems, architecture, planning, art, and their combined ability to improve the quality of our lives and environment. I can also vividly remember the nagging feeling that the human condition—particularly in the urban centers—was not high on the academic agenda in design schools. Our exposure to public spaces was European parks and plazas; Central Park and its offshoots; and barren, lifeless modernist plazas. A few of us in school had already committed our professional futures to the cities. The potential to touch millions of ordinary people was obvious to me and irresistible, but nowhere in my academic experience was there mention of humanism, human behavior, sociology, or psychology.

Fortunately, early years of practice in New York City introduced me to *The Social Life of Urban Spaces* and its author. Whyte's teachings provided the missing link between my artistic sensibilities and the principles of public open space design and management. Over time—and with the benefit of his direct consultations and critiques—I have been able to design, observe, learn, and improve my work in ways that have miraculously transformed neighborhoods and cities. Each new park or plaza design commission follows an evolutionary process that explores new ways in which we can artistically express our time and culture, guided by Holly's principles and gentle whispers from just over my shoulder.

The Wit and Wisdom of Holly Whyte

Gathered by Albert LaFarge

- People sit most where there are places to sit.

- Good aesthetics is good economics.

- What attracts people most, it would seem, is other people.

- The street is the river of life of the city; and what is a river for if not to be swum in and drunk from?

- The human backside is a dimension architects seem to have forgotten.

- New York is a city of skilled pedestrians.

- Supply creates demand. A good new space builds its constituency—gets people into new habits, like eating outdoors; induces them to new paths.

- So-called undesirables are not the problem. It is the measures taken to combat them that is the problem. . . . The best way to handle the problem of undesirables is to make a place attractive to everyone else.

- Most ledges are inherently sittable, but with a little ingenuity and additional expense they can be made unsittable.

- It is difficult to design a space that will not attract people. What is remarkable is how often this has been accomplished.

- Walls are put up in the mistaken notion that they will make a space feel safer. Too often they make it feel isolated and gloomy.

- By default street vendors have become the caterers of the city's outdoor life. They flourish because they are servicing a demand the downtown establishment does not.

- When people start to fill up a space, they do not distribute themselves evenly across it. They go where the other people are. Dense areas get denser.

- Planners sometimes worry that a place might be made too attractive and thereby overcrowded. The worry should be in the opposite direction. The carrying capacity of most urban spaces is far above the use that is made of them.

- Simulated cities for people who don't like cities, it turns out, are not such a good idea after all.

- Blank walls proclaim the power of the institution and the inconsequence of the individual, whom they are clearly meant to intimidate. Stand by the new FBI headquarters in Washington. You feel guilty just looking at it.

- In the matter of zoning bonuses and incentives, what you do not specify you do not get.

- In some American cities so much of downtown has been cleared for parking that there is now more parking than there is city. . . . One of the greatest boons of mass transit is what it makes unnecessary: the leveling of downtown for parking.

- Food attracts people who attract more people.

- Big buildings cast big shadows. Bigger buildings cast bigger shadows.

- People in big cities walk faster than people in smaller cities.

- The waterwall in Greenacre Park makes fine music.

- In almost every U.S. city the bulk of the right of way is given to vehicles; the least, to people on foot. This is in inverse relationship to need.

- Ninety-fourth Street is the honkingest street in town. I love it. I live here.

PART TWO

From City Parks
to Regional Green Infrastructure

As access to "country" beyond metropolitan areas gets ever more distant
and frustrating, existing parks and other preserved greenspaces within reach of
the four-fifths of Americans who live in metro areas become increasingly vital.
Accordingly, Part II addresses one of William Whyte's favorite topics: city parks
and regional greenspaces. Who better to open this section of the book than Peter
Harnik, one of the founders of the Rails to Trails Conservancy and now director
of the Green Cities Program based at the Washington, D.C., office of the Trust
for Public Land. Harnik's essay is based on his seminal research on the nuts
and bolts (e.g., design, management, finance) of urban park systems across the
United States.

Robert L. Ryan, professor of landscape architecture at the University of Massachusetts Amherst, complements Harnik's broad overview with his essay on
how local residents may "adopt" parks in their vicinity, thereby helping maintain
the greenspace itself while emotionally "bonding" with the park as part of their
daily urban living experience. Thus, parks may contribute to local "sense of place"
as the focus of maintenance and improvement efforts that in turn bring people
into enjoyable contact with one another (a cardinal Whyte principle).

Michael C. Houck is a key "mover and shaker" in Portland, Oregon's ongoing
quest to preserve and extend one of the nation's best-known regional greenspace
systems. At the Portland Audubon Society since 1982 and more recently through
his Urban Greenspaces Institute, Houck champions a wide spectrum of initiatives
to save farmland, restore streams and wetlands, protect endangered species
habitat, and expand existing parks and greenways. As his contribution demonstrates, Houck is a consummate networker with a strong grounding in the historical and natural science contexts of regionalism in Portland and elsewhere.

Urban parks and greenspaces in Whyte's day were traditionally designed to
foster sedentary relaxation amid aesthetic surroundings, with active recreation
largely confined to playgrounds and athletic fields for the young and fit. In light of
today's obesity crisis, urban open spaces must provide opportunities for vigorous
outdoor recreation and physical fitness for an aging and culturally diverse public.
Anne Lusk, an experienced public health researcher, offers practical advice on
the design of urban greenspaces and linear corridors to encourage such activities
as running, cycling, skating, tennis, rock climbing, and other energetic outdoor
pastimes.

Geographers William D. Solecki and Cynthia Rosenzweig are longtime collaborators studying the actual and potential effects of environmental change in the
New York metropolitan region. Scarcely as cohesive as the Portland, Oregon,

region (or Chicago for that matter), New York City nevertheless has a long tradition of "large-vision" urban greenspace plans dating back to Olmsted's Central
Park. Since 1928, the Manhattan-based Regional Plan Association (RPA) has
tried to foster a broad functional and perceptual sense of "regionalism" within its
tristate, thirty-one-county planning area. RPA's "Regional Greensward Plan" and
its "H_2O: Highlands to Ocean" program (Hiss and Meier 2004) represent the lat-
est chapter of this chronicle. Solecki and Rosenzweig jump one step further with
their concept of a New York Urban Biosphere Reserve in relation to the UNESCO
international network of biosphere reserves.

Reference

Hiss, T., and C. Meier. 2004. *H_2O: Highlands to ocean.* Morristown, NJ: Geraldine R. Dodge Foundation.

Among their many functions, parks in cities serve as gathering spaces where people may participate in shared civic experience including demonstrations, celebrations, and grieving. (*Top*) A gathering in Jackson Park on the Chicago lakefront to protest the Vietnam War, circa 1970 (a NIKE anti-aircraft battery was hidden behind the chain-link fence). (*Bottom*) The Vietnam Memorial on the Mall in Washington, D.C. (Photos by R. H. Platt.)

The Excellent City Park System

What Makes It Great and How to Get There

Peter Harnik

The Changing Roles of City Parks

The total area covered by urban parkland in the United States has never been counted, but it certainly exceeds one million acres. The fifty largest cities (not including their suburbs) alone contain more than 600,000 acres, with parks ranging in size from the jewellike 1.7-acre Post Office Square in Boston to the gargantuan 24,000-acre Franklin Mountain State Park in El Paso. The exact number of annual visitors has not been calculated either, but it is known that the most popular major parks, such as Lincoln Park in Chicago and Griffith Park in Los Angeles, receive upwards of twelve million users each year, while as many as twenty-five million visits are made to New York's Central Park annually, more than the total number of tourists coming annually to Washington, D.C.

What makes a park system "excellent"? From the very beginnings of the urban parks movement, dating back to the Olmsted parks of the second half of the nineteenth century, there has been interest in this question. At first, attention was focused on individual parks; later the inquiry was expanded to what constitutes greatness for a whole network. Each analysis, however, was confined to a limited view of parks, looking at an isolated factor such as location, size, shape, plantings, uses, or historical integrity. No analysis addressed the creation of a *park system* as a singular entity within a city infrastructure.

Today's older city park systems are being revisited in light of twenty-first-century demands and demographics; likewise many cities, counties, and regional park authorities and environmental organizations are piecing together new kinds of parks and systems of greenspaces under such rubrics as greenways, conservation areas, areas of special significance (ecological, geological, or cultural), and environmental education centers (figure 1). With 80 percent of people in the United States now living in metropolitan areas, there is renewed interest in understanding more precisely the relationship between cities and the open space within them. What factors lead to all-around park excellence?

Beginning in 1859, when Frederick Law Olmsted, Calvert Vaux, and more than

This essay is based on a publication of the same title published by the Trust for Public Land, copyright 2003.

Figure 1   New use for an old fountain: Bryant Park in New York. (Photo by R. H. Platt.)

three thousand laborers created Central Park, a wave of enthusiasm for urban "pleasure grounds" swept the nation. Thousands of parks were constructed and millions of words were written about their features and attributes. During the height of the city park movement, from about 1890 to 1940, great efforts were made to plan for parkland, to understand the relationship between parks and surrounding neighborhoods, and to measure the effect of parks. Leaders in Boston, Buffalo, Seattle, Portland (Oregon), Denver, Baltimore, and elsewhere proudly and competitively labored to convert their cities from drab, polluted industrial cores into beautiful, culturally uplifting centers. They believed that a well-designed and maintained park system was integral to their mission.

Inspired by boulevard systems in Minneapolis and Kansas City, and by Olmsted's "Emerald Necklace" in Boston, many cities sketched out interconnected greenways linking neighborhoods, parks, and natural areas. Careful measurements were made of the location of parks and the travel distance (by foot, generally) for each neighborhood and resident. The field of park research was supported by the federal government through the National Conference on Outdoor Recreation, which provided funding for data collection, research, analysis, and dissemination. During the early twentieth century, the purpose and design of parks metamorphosed, but these areas remained so important to cities that even during the Great Depression many park systems received large influxes of money and attention through the federal government's relief and conservation programs. A case in point was the renovation of the New York City parks under the direction of Robert Moses as City Parks commissioner during the 1930s (Caro 1986).

After World War II, the nation's attention turned toward the development of suburbs, and commitment to the parks and public spaces of cities began to wane. There was even a naïve assumption that private suburban backyards could replace most of the services provided by public city parks. Many of the ideas regarding the role of parks in city planning and community socialization were lost. More important, ideas about measuring park success, assuring equity, and meeting the needs of changing users languished. Between 1950 and 2000, many of the nation's urban park systems fell on hard times. Few cities provided adequate maintenance staffing and budgets, and most deferred critically needed capital investment. Many parks suffered from overuse, revealing trampled plants and grass, deteriorated equipment, erosion, and loss of soil resiliency and health. Others declined from underuse and resulting graffiti, vandalism, invasion of noxious weeds, theft of plant resources, and crime.

The decline was camouflaged. In the older northern cities, general urban deterioration grabbed headlines and made parks seem of secondary importance. In the new cities of the South and West, low-density development made parks seem superfluous. Intellectual inquiry into traditional city greenspace dwindled to almost nothing. An exception to this dismal trend beginning in the 1960s was a growing interest in "urban ecology" and the value of restoring and preserving wetlands, deserts, forests, and grasslands within and near cities for their ecological values and benefits (e.g., McHarg 1968; Spirn 1984; Platt, Rowntree, and Muick 1994).

Every pendulum eventually swings back, and the effort to revive city park systems has slowly gained momentum. When the Trust for Public Land (TPL) was founded in 1972, it was the first national conservation organization with an explicit urban component to its work. At the same time, fledgling neighborhood groups began forming to save particular parks, either through private fundraising or through public political action. There arose a new appreciation of the genius and work of Frederick Law Olmsted, and in 1980 the Central Park Conservancy was founded. In that same year, pioneering research by William H. Whyte resulted in the publication of *The Social Life of Small Urban Spaces* (1980) and the formation of the Project for Public Spaces, Inc. in New York City. The rise of the urban community gardening movement and the spread of park activism to other cities led in 1994 to a $12 million commitment by the Lila Wallace–Reader's Digest Foundation and the creation of the Urban Parks Institute and the City Parks Forum. Meanwhile, city park directors formed their own loose network through the Urban Parks and Recreation Alliance.

Beginning around 1995, many older cities such as Chicago, Boston, Washington, D.C., and Cleveland started to bounce back from years of population loss and fiscal decline, partly owing to "gentrification," the return of suburban empty nesters and young professionals to restore older urban neighborhoods (with consequent displacement of the low-income households occupying them). With new residents and a greater sense of optimism, these cities and other places like them began to seek to reestablish a competitive edge by reviving and expanding their municipal assets such as parks, museums, sports stadiums, and performing arts centers. Elsewhere, in fast-growing Sunbelt cities such as Charlotte, Dallas, and Phoenix, planners were belatedly trying to create vibrant downtowns and walkable neighborhoods for a more cohesive urban identity. In both old cities and new, there is rising interest in the use of parks to promote urban vitality (figure 2), an interest that has been encouraged by the smart growth and New Urbanist movements since the mid-1990s.

By the mid-1990s, after years of creating parks, TPL became concerned about the woeful lack of basic information about city systems. TPL initiated a research program to collect data and revisit old ideas about parks and cities. Statistics regarding land ownership, recreational facilities, and budgets were assembled for the first time in more than fifty years. This research led to the book *Inside City Parks*

Figure 2   Old use for a new building: Frank Gehry's outdoor concert venue at Millennium Park in Chicago. (Photo by R. H. Platt.)

copublished by the Urban Land Institute and TPL (Harnik 2000), which examined and compared the park systems of the twenty-five largest U.S. cities. The book generated a storm of publicity for places given the highest and lowest rankings and also stimulated leaders of many other cities to ask to be included in future studies. At the same time, a number of correspondents suggested that the research was too restricted. The breadth and depth of a park system, they said, cannot be determined by simple statistics on acreage, recreation facilities, and budgets. It was time to determine exactly what factors make for a truly excellent city park system.

To study this question—"What makes an excellent city park system?"—TPL convened a multidisciplinary group of twenty-five urban and park experts for an intensive two-day meeting in Houston in October 2001. This workshop yielded a list of seven broad measures that make the greatest difference in defining a successful system. TPL's goal for this project is to re-create the kind of framework that existed in the early part of the twentieth century to sustain city parks as valued components of a vital urban community

Seven Measures of an Excellent City Park System

1. Mission Statement and Updates

Park systems do not just "happen." Wild areas do not automatically protect themselves from development, outmoded waterfronts do not spontaneously sprout flowers and promenades, and flat ground does not morph into ball fields. Even trees and flora of the desired species do not spontaneously grow in the right places. Interested citizens must identify the goals of the park system, including functions to be served, management, and landscaping. The parks department must then use that mandate as a basis for its mission statement and the definition of its core services.

Most big-city park agencies have a legislative mandate and a mission statement, but about 20 percent of them have not formally defined their core services. A failure to develop this definition and to check periodically whether it is being followed can lead to departmental drifts due to political, financial, or administrative pressures. Having a strong concept of mission and core services, on the other hand, can stave off pressures to drop activities or add inappropriate tasks.

To inform the public, the department should regularly publish an annual report summarizing its system and programs and showing how well it fulfilled its mandate. Less than half of big-city agencies publish an annual report, and most of the reports provide "soft" concepts and images rather than precise information, such as number of activities held, number of people served, and other specific outcomes and measurable benefits. Few agencies give a comprehensive budgetary report, and fewer still look honestly at challenges that were not adequately met and how they could be tackled better in the future.

2. Ongoing Planning and Community Involvement

To be successful, a city park system needs a master plan. A plan is more than an "intention"; it is a document built on a process, demonstrating a path of achievement, and expressing a final outcome. The department's master plan should be substantiated thoroughly, reviewed regularly, and updated every five years. The agency should have a robust, formalized community involvement mechanism, which means more than posting the document on a website and hoping for feedback. The ideal master plan should have at least the following elements:

- An inventory of natural, recreational, historical, and cultural resources
- A needs analysis
- An analysis of connectivity and gaps
- An analysis of the agency's ability to carry out its mandate
- An implementation strategy (with dates), including a description of the roles of other park and recreation providers
- A budget for both capital and operating expenses
- A mechanism for annual evaluation of the plan.

Philadelphia Green and Philadelphia Parks Alliance

There may or may not be brotherly love in Philadelphia, but there certainly is love of parks. The city has 138 "friends of parks" organizations: two of them operating on a citywide basis and the rest focusing on one particular park or playground.

The largest organization is Philadelphia Green, a division of the venerable Pennsylvania Horticultural Society, which began in 1974 as a community vegetable gardening project and today is an urban greening powerhouse with a staff of twenty-eight and a budget of $4 million. Philly Green partners with private and public groups to landscape and maintain public spaces downtown and along gateways such as the road to the airport, but the main thrust of its work is in neighborhoods. There the multipronged program is growing crops, instilling pride, teaching skills, developing microbusinesses, stopping illegal dumping on vacant lots, refurbishing parks, and stimulating the redevelopment of blighted neighborhoods. Twice a year, the group organizes massive cleanups called "Spring into Your Park" and "Fall for Your Park." All told, Philly Green has helped plan and implement more than 2,500 greening projects in the city.

The other citywide organization, Philadelphia Parks Alliance (PPA), is more explicitly advocacy-oriented, pushing for more funding and for better stewardship of the large Fairmount Park system. Formed in 1983 by Sierra Club activists, the group incorporated separately and now has a $300,000 budget and a staff of three. With a quarterly newsletter, annual meetings that include many of the local park groups around the city, and a "Green Alert" mailing list of 550 leaders, PPA is at the center of the campaign it calls "A New Era for Philadelphia's Parks." Some 136 organizations help maintain their local parks by removing trash, programming activities, helping with special projects, organizing celebrations, watching out for problems, and showing up at City Hall every year at budget hearing time.

Although five years may seem a short lifespan for a plan, it is startling to realize how rapidly urban circumstances change. In TPL's survey, about two-thirds of agencies surveyed were operating under out-of-date master plans, and some were relying on plans formulated before the rise of computers and geographic information systems, not to mention dog parks, mountain bikes, ultimate Frisbee, girls' soccer leagues, skateboard courses, and cancer survivors' gardens, among other innovations.

The ability of good planning to build community support was demonstrated recently in Nashville where in 2002 Mayor Bill Purcell initiated a yearlong parks and greenways process, the first such citywide conversation in the one-hundred-year history of its parks. Upon completion, resident support had been so solidified that the city council enthusiastically funded a $35 million capital spending plan, the largest Nashville park appropriation ever.

Although most park agencies have plans, too often they never reach fruition because key elements are trumped by other agencies or private interests. Visions of a new waterfront park, for instance, may be for naught if the transportation department has its own designs on the same parcel. Any park plan (and its implementation strategy) should be coordinated with plans for neighborhoods, housing, tourism, transportation, water management, economic development, and education and health, among other factors. Ideally, the agencies will reach agreement; if not, the issue should go to the mayor or city council for resolution, with plenty of public involvement and support from propark advocates.

As confirmation of its involvement with the community, the department should have formal relationships with nonprofit conservation and service-provider organizations. These arrangements may or may not involve the exchange of money, but they should be written down explicitly and signed, with clear expectations, accountability, and a time limit that requires regular renewal. Having formal relationships not only enables a higher level of service through public-private partnership; it also provides the agency with stronger private-sector political support if and when it is needed.

Finally, no city can have a great park system without a strong network of park "friends" groups, private organizations that serve as both supporters and watchdogs of the department. Ideally, a city will have one or two organizations with a full citywide orientation, assuring that the system as a whole is well run and successful, and also scores of groups that focus on an individual park and its surrounding neighborhood, concentrating on everything from cleanliness, safety, and quality to programming, signage, and special fund-raising.

3. Sufficient Resources to Meet the System's Goals

Obviously, a park system requires a land base. Yet the size of that base is not an immutable number: big-city systems range in size from almost 20 percent of a city's area down to 2.5 percent, and from more than 45 acres per one thousand residents to just over 3 acres per thousand. Although there is no ordained "optimum" size, a city's system should be large enough to meet the goals outlined in its mission statement and master plan.

Despite the truism "If you don't measure, you can't manage," many cities do not have accurate figures on their systems. Every agency needs to know the extent of its natural and historical resources—land, flora, buildings, artwork, waterways, paths, roads, and much more—and have a plan to manage them sustainably. It is important to publish these numbers annually to track the growth (or shrinkage) of the system over time. Ideally, the agency should be able to place a financial value on its holdings and should have a plan to pay for replacing every structure in the system.

Because it is so much more expensive to create and operate "designed" landscapes (constructed parks that are mowed or regularly cleaned up) than natural landscapes (those that are left alone, except for the occasional trail), it is valuable to know the system's allocation between these two categories, both actual and planned. The TPL survey reveals a large range: some urban park agencies consist entirely of designed lands and no natural properties at all, whereas others have as little as 10 percent designed and 90 percent natural.

Newer systems in younger cities have more potential for expansion of parks' area than older systems in mature, nonexpanding cities, although older cities can nevertheless increase the size of their park systems as well. Since the 1970s, for instance, the amount of parkland in Denver and Seattle grew by more than 44 percent each. Conversely, some "new cities" have fallen behind in the effort to add parkland. Even though the Colorado Springs park system grew in acreage by 185 percent between 1970 and 2002, the city itself grew even more—206 percent during the same time—yielding a slight net loss over the period.

Even cities that are considered "all built out" can use redevelopment to increase parkland. Outmoded facilities like closed shipyards, underutilized rail depots, abandoned factories, decommissioned military bases, and filled landfills can be converted to parks. Sunken highways and railroad tracks can be decked over with parkland. Denver is even hard at work depaving its old airport to restore the original land contours and create what will be the city's largest park.

In New York City, the Department of Parks and Recreation collaborated with the Department of Transportation to convert 2,008 asphalt traffic triangles and paved medians into "greenstreets" or pocket parks and tree-lined malls that are then maintained by community residents and businesses. In other cities, school systems and park departments are breaking down historic bureaucratic barriers and signing joint use agreements to make school athletic fields available for neighborhood use after school hours.

In addition to land, the parks and recreation department needs, of course,

The Chicago Park District: Increasing Landholdings, Assuring Revenue

Despite its world-famous lakefront system, Chicago has a shortage of parkland in the rest of the city. Under the leadership of Mayor Richard M. Daley, however, the metropolis has embarked on an ambitious and thoughtful effort to acquire additional land to more equitably serve its residents. Called the CitySpace Plan, it is a joint program of the Chicago Planning Department, the Chicago Park District, the Forest Preserve District of Cook County, and the Chicago Public Schools.

Finding that 63 percent of Chicagoans lived in neighborhoods where parks are either too crowded or too far away, CitySpace in 1993 set out to methodically gain open space in five ways:

- Convert asphalt schoolyards and portions of school parking lots to grass fields
- Create trails, greenways, and wildlife habitat alongside inland waterways such as the Chicago River and Lake Calumet
- Convert vacant, tax-delinquent private lots into community gardens
- Redevelop abandoned factories into mixed-use developments that include parkland
- Build parks on decks over rail yards.

Before plunging into this formidable task, the planners carried out a detailed analysis of virtually every square foot of the city, identifying both community needs and each parcel of public and private open space. They also worked with more than a hundred other government agencies and civic, community, and business organizations to reach a full understanding of the many economic and regulatory processes that tend to stimulate (or prevent) the creation of parkland. By the end of the study, the CitySpace team was able to use the complexity of Chicago's bureaucracy to its advantage instead of being stymied by it. Among the action steps developed were specific strategies to acquire funding, to obtain abandoned, tax-delinquent properties, to mandate open space in special redevelopment zones, and to change zoning laws.

The outcome has been impressive. Since 1993, under guidance of the plan, Chicago has added 99 acres to its park system, 150 acres to its school campus park network, a 183-acre prairie for a future state open space reserve, and two miles of privately owned but publicly accessible riverfront promenade. The city has also leased ten acres along the Chicago River and provided permanent protection of forty community gardens. The total cost of this increase has been in excess of $30 million.

One reason the Chicago Park District has been able to afford land acquisition in a staggeringly expensive market is that the agency is authorized to receive a portion of the city's property tax. This guaranteed source of revenue not only shields the Park District from city council politics and cutbacks, but it also enables the agency to issue bonds because lenders know that repayment is guaranteed from tax revenue.

"The CitySpace Plan enabled us to focus our acquisitions in the geographical areas of need," said Bob Megquier, director of planning and development for the Park District. "It may be a slow and costly process, but at least we know that we are putting our resources in the right places."

Only a handful of other city park agencies have a charter that mandates receipt of a portion of the property tax, and most of them are among the better-funded departments. Chicago Park District, for instance, spends $123 per resident, more than all but four of the big-city park agencies.

sufficient public revenue for land management and programs. Such funds entail both an adequate operating budget and a regular infusion of capital funds for major construction, repairs, and land acquisition. A detailed survey of the fifty-five largest cities showed that in fiscal year 2000, the "adjusted park budget"—the amount spent by each city on parks operations and capital, minus everything spent on such big-ticket items as zoos, museums, aquariums, or planetariums—came to an average of $79 per resident. Although that figure is probably not high enough considering that every system is far behind its needs, in current dollars this figure may be considered a minimum level.

Moreover, there should be an effective, complementary private fund-raising effort, one that serves not only signature parks but also the whole system. Although private efforts should never be designed to let the local government "off the hook," they can be valuable in undertaking monumental projects or in raising work to levels of beauty and extravagance that government on its own cannot afford. Private campaigns are also effective in mobilizing the generosity of corporations, foundations, and wealthy individuals who otherwise would not contribute to government agencies.

Excellent park departments not only receive adequate funding, but also spend

---

Phoenix: A High Level of Stewardship

"Stewardship" involves land, money, planning, public participation, commitment, awareness, and volunteerism. Phoenix represents excellence in stewardship.

The Phoenix Parks and Recreation Department starts with an excellent planning process during which it inventories resources and plans how to protect them, analyzes geographical and user needs, reviews gaps in the system's connectivity, and sets forth budgets and an implementation strategy.

Through good fortune and good skills, the agency has been allotted a generous budget that allows it to maintain a large staff, including more than forty foresters, horticulturalists, and landscape architects, to assure good planning and nature management. The agency's maintenance budget amounts to more than $11,000 for every acre of "developed" parkland, a very high level.

Volunteerism is also strong in Phoenix parks. In 2001, more than 22,000 volunteers donated more than 200,000 hours of work. In addition, there is a private Phoenix Parks and Conservation Foundation through which citizens and businesses can make donations for specific projects. Past efforts have included the Japanese Friendship Garden, the Irish Cultural Center, and a cancer survivors' park. The foundation recently assumed the role of a land trust, holding land donations and receiving mitigation funds on behalf of the parks department from such agencies as the U.S. Army Corps of Engineers.

In 2001, Phoenix ranked first overall in a comprehensive national study that measured how well U.S. cities deliver government services to local citizens. The Phoenix parks department ranked at the top of its class also.

their money wisely and commit themselves to effective stewardship. Outstanding stewardship means having enough qualified natural resources professionals to properly oversee the system and manage the work of pruners, mowers, and other laborers. Moreover, because a system rarely has enough paid staff to accomplish all its goals, the excellent park department has a high-visibility, citizen-friendly marketing program whereby members of the public can understand the stewardship of the system and become involved, if they wish.

Finally, park departments must track their expenditures accurately and be able to report them to the public usefully and understandably. Most agencies have the raw information but too many of them do not provide it; numbers are either difficult for politicians, reporters, and the general public to obtain or the statistics are put forth incomprehensibly.

4. Equitable Access

The excellent city park system is accessible to everyone regardless of residence, physical abilities, or financial resources. Parks should be easily reachable from every neighborhood, usable by those who are handicapped or challenged, and available to low-income residents.

Most cities have one or more very large unspoiled natural areas. By virtue of topography—mountain, wetland, canyon, stream valley—they are not, of course, equidistant from all city residents. But *created* parks—squares, plazas, playgrounds, neighborhood parks, ball fields, linear greenways—should be sited in such a way that every neighborhood and every resident are equitably served. Preferably, people and parks are no farther than ten minutes apart by foot in dense areas or ten minutes apart by bicycle in spread-out sections. Moreover, it is not enough to measure access purely from a map; planners must take into account such significant physical barriers as uncrossable highways, streams and railroad corridors, or heavily trafficked roads. Also, the standard for acceptable distance should not be based on an idealized healthy adult, but rather on a senior citizen with a cane, a parent pushing a stroller, or an eight-year-old riding a bicycle. Unfortunately, the TPL survey found that most cities do not know how many residents live unreasonably far from a park.

Cities should also ensure park access by a wide range of challenged persons, including those who are elderly, infirm, blind, or confined to a wheelchair. Access includes, for example, appropriate surfacing materials, ramps, signs, and handicapped parking.

Finally, agencies must ensure equitable access for those who cannot pay full price. Although it is acceptable to charge appropriate fees for some park facilities and programs, agencies should consciously plan for the approximately 20 percent of residents who cannot afford such fees, using such alternatives as scholarships, fee-free hours, fee-free days, or sweat-equity volunteer work.

Denver Parks: No More than Six Blocks from a Park

In Denver, more than nine out of ten residents live within six blocks of a park. This statistic is impressive not only because of the accessibility that it represents but also because Denver has obtained such data. "Geography is everything," explains Susan Baird, manager of the Master Plan Process for Denver Parks and Recreation. With park access as the project's focus, Baird worked with consultants on a geographic information system analysis that went beyond a neighborhood analysis all the way to a building-by-building study. Researchers used a computer model to draw a six-block-radius circle around each traditional park or protected natural area. They did not count any of the city's numerous parkways, maintaining that although the parkways are visual amenities, they are not directly usable as parks.

According to Baird, "The goal wasn't just any six blocks. We said that it needs to be a *walkable* six blocks, meaning that people can get to the park without having to cross a highway, railroad track, or body of water. Crossing a six-lane road is not access." Thus, the Denver team truncated circles wherever they crossed barriers, further clarifying which residents did not have good enough access. Funding for the analysis came from capital appropriations for the master plan.

At eleven acres per one thousand residents, the total amount of parkland in Denver is not extraordinarily high, primarily because the city does not have any huge parks comparable to those in Philadelphia, Kansas City, Los Angeles, and many other places. Denver, however, more than compensates for size with distribution. It is also committed to improvement: Denver Parks and Recreation hopes to tighten the radius down to four blocks, or about one-third of a mile once the six-block criterion is achieved.

5. User Satisfaction

By definition, the excellent city park system is well used. Having high usership is the ultimate validation that it is attractive and that it meets people's needs. High attendance also increases safety because there are more "eyes on the park." (See the essay by Robert L. Ryan in this volume.)

Knowing the level of park use requires measuring it, not only for an estimate of a gross total but also to identify users by location, time of day, activity, and demographics. In addition, finding out the satisfaction level requires asking questions, not only of users but of nonusers as well. These efforts must be carried out on a continuing basis using standardized methodology to discern trends over time.

The TPL study found that an overwhelming number of city park agencies are unaware of their parks' total usership. Not having this number severely reduces an agency's ability to budget and to request adequate funding from the city council. Most departments can track their paying users, such as golfers playing rounds, swimmers using pools, and teams renting fields, but those users are only a tiny fraction of the true total. The lack of basic information is in stark contrast to, for instance, the transportation department, the school system, or the welfare department, which can all make strong, factual cases to justify their budget requests.

As for satisfaction, most agencies rely on informal feedback such as letters of complaint or messages relayed back by the staff. This process is unbalanced and ineffective, and it does not provide the agency with clear direction. It therefore tends to result in a park system that meets the efficiency needs of the provider rather than the comfort needs of the user. (For instance, some park agencies "solve" the problem of dirty bathrooms not by cleaning but by permanently locking them.)

It is difficult to count all passive users of a system accurately. Repeated observation, selective counts, and extrapolations over time, however, can provide meaningful data. Chicago takes aerial photos of large events and then uses a grid to count participants. The city also sets up electronic counters to measure the number of users passing a given point.

6. Safety from Physical Hazards and Crime

To be successful, a city park system should be safe: free both of crime and of unreasonable physical hazards, from sidewalk potholes to rotten branches overhead. Park departments should have mechanisms to avoid and eliminate physical hazards as well as ways for citizens to report problems easily.

Crime, of course, is dependent on a large number of factors that are beyond the reach of the park and recreation department, such as poverty, drug and alcohol use, population demographics, and lack of stabilizing neighborhood institutions. Yet the park agency has some control over other factors, including park location, park design, presence of uniformed personnel, presence of park amenities, and availability of youth programming. Ultimately, the greatest deterrent to crime is the presence of large numbers of users.

Park visitors are also reassured if they see uniformed employees. Even if the number of actual police or rangers is quite small and their rounds infrequent, the perception of order and agency responsibility can be extended simply by dressing all park workers and outdoor maintenance staff in uniform.

Similarly, well-run youth recreation programs have been shown to decrease delinquency and vandalism. Austin, Texas, for instance, created what it called the Social Fabric Initiative, a multilayered program that includes a summer teen recreation academy, a neighborhood teen program, an art-based program called "Totally Cool, Totally Art," and a roving leader program that sends trained staff into neighborhoods with vans, sports equipment, and art projects. The excellent park system takes it even further by tracking youth crime by neighborhood over time.

Because parks and their surrounding neighborhoods are interrelated, basic to any safety strategy is the accurate, regular collection of crime data within as well as in nearby neighborhoods. (Only about half the surveyed agencies currently collect such data and, of those that do, most have no strategy to use the information.) Another valuable piece of information is the ratio of male to female users in each park because a low rate of female users may indicate that the park feels unsafe.

7. External Benefits of Parks to the City

Benefits of a park system should extend beyond the boundaries of the parks themselves. In fact, the excellent city park system is a form of "natural infrastructure" that provides many "ecological services" (Daily 1999) to the city as a whole:

- Cleaner air, as trees and vegetation filter out pollutants
- Moderation of microclimate and reduction of the "urban heat island"
- Cleaner water, as roots trap silt and contaminants before they flow into local water bodies
- Reduced health costs through opportunities for physical fitness
- Improved learning opportunities from "outdoor classrooms"
- Increased urban tourism with resulting increased commerce and sales tax revenue
- Increased business vitality based on attraction of good parks
- Natural beauty and respite from traffic and noise.

City parks do not exist in a vacuum. Every city is a complex and intricate interplay between the private space of homes and offices, the semipublic spaces of shops, and the fully public space of parks, plazas, streets, preserves, and natural areas. The goals are a park system that enriches cities and cities that nourish their parks.

References

Caro, R. 1986. *The power broker.* New York: Knopf.

Daily, G., ed. 1999. *Nature's services.* Washington, DC: Island Press.

Harnik, P. 2000. *Inside city parks.* Washington, DC: Urban Land Institute and Trust for Public Land.

McHarg, I. 1968. *Design with nature.* Garden City: Doubleday.

Platt, R. H., R. Rowntree, and P. Muick 1994. *The ecological city: Restoring and preserving urban biodiversity.* Amherst: University of Massachusetts Press.

Spirn, A. W. 1984. *The granite garden: Urban nature and human design.* New York: Basic Books.

Whyte, W. H. 1980. *The social life of small urban spaces.* Washington, DC: The Conservation Foundation.

The Role of Place Attachment in Sustaining Urban Parks

Robert L. Ryan

Sustaining urban parks requires developing a constituency of dedicated park users, neighbors, and stewards. Urban parks that do not have a cadre of local residents who have "adopted" them are subject to vandalism, neglect, and even destruction. Yet there are strategies for planning, designing, and managing parks in a manner that builds an attachment between people and their parks. Several research studies on urban parks and natural areas illustrate the factors that influence people's attachment to these precious urban natural areas. An important part of this work is to expand the definition of traditional park users, as studied by William H. Whyte (1980, 1988), to a broader group of concerned citizens, including those who live and work near urban parks, volunteer stewards, and even those who simply pass by these green spaces on their way to work or home.

The goal of this essay is to help park planners, managers, and advocates create successful urban parks and open spaces by fostering an attachment between urban residents and their parks. Expanding the definition of urban parks is an important part of this effort. In addition, the following key questions are addressed here:

- What factors might lead people to develop an attachment to an urban park?
- How can park managers, advocates, and planners nurture, understand, and respect this relationship between people and parks?
- What strategies might be useful for building an attachment between people and parks?
- How can park managers develop parks that serve a diverse set of park users and avoid the domination of park use by a particular set of users (e.g., drug traffickers, teens, or dog walkers).

Benefits of Urban Parks

Urban parks and open spaces are essential for the ecological health of urban environments (Platt, Rowntree, and Muick 1994). These urban greenspaces include traditional parks as well as other public greenspaces such as nature preserves, plazas, and cemeteries. "The Humane Metropolis" relies on its city and regional parks to provide vital ecological benefits, including cleaning air and water systems, cooling the urban heat island, and providing wildlife habitat (Spirn 1984; Hough 1994). It is estimated that even densely populated New York City retains 27 percent of its land, or approximately 17,000 acres, as parks and open space, and the majority of this land is in a "natural" undeveloped condition as forests and wetlands (Benepe 2002).

For many urban residents, however, parks and open space provide much more than environmental benefits. Parks are perceived as an essential part of the quality of life in densely populated urban areas (Harnik 2000). From the very beginning, urban parks were designed for human leisure and recreation (Hough 1994). Whyte documented the importance of smaller parks and plazas to urban dwellers in his groundbreaking book, *The Social Life of Small Urban Spaces* (1980). Urban parks and other greenspaces provide restoration from the mental fatigue caused by modern urban life (Kaplan, Kaplan, and Ryan 1998). The psychological benefits of urban greenspace have just begun to be explored during the past few decades. Urban parks and trees have a special importance to urban residents. People have a strong attraction for urban parks and trees (Dwyer, Schroeder, and Gobster 1994). Moreover, people may develop an emotional attachment to urban parks and natural areas, with profound implications for the design and management of these areas (Ryan 1997, 2000, 2005).

This strong appreciation for urban parks and other conservation land has prompted a groundswell of public support for land acquisition in the face of ever-expanding urban sprawl. In 2001, voters across the United States approved more than $1.7 billion in new conservation funding (Trust for Public Land 2002). The public's appreciation for parks and open space has also been manifested in a proliferation of private nonprofit park conservancies, land trusts, and foundations that raise money for park maintenance and improvement. For example, the Central Park Conservancy has taken over management of the park from the City of New York and raises an estimated $15 to $20 million annually for park maintenance and renovation (Benepe 2002). The public has also responded by volunteering time to park stewardship programs. It is estimated that the more than 70,000 volunteers in New York City Parks donated an estimated one million hours in service (Benepe 2002).

Unfortunately, not all urban parks are well loved or cared for; many urban parks are neglected, forgotten places. As of December 2002, the New York City Parks Department had reduced its staff to two thousand, compared with six thousand employees in 1970, in the face of budget cuts, and this dilemma was shared by many other city park systems (Lutz 2002). Underused and underfunded, parks become dangerous places that urban residents fear, continuing the cycle of neglect. For urban parks to become sustainable, they need a group of dedicated citizens (park users, neighbors, and volunteer stewards) who are willing to protect, nurture, and advocate for them.

Urban Park Studies: Creating Measures of Success

At the heart of successful park planning, design, and management is an understanding of what the public wants in its urban parks. As noted by Whyte (1988, 109), however, the public is often overlooked when parks are designed: "It is difficult to design a space that will not attract people. What is remarkable is how often this has been accomplished." Whyte used behavioral observation of people in urban parks and plazas in New York City to understand the factors that constitute a successful urban plaza. He focused primarily on the physical features within these spaces. He observed that those plazas that provided ample seating and, if possible, movable chairs that allowed people to create their own seating arrangements were the most heavily used. Other important factors included creating a comfortable microclimate, such as sunny areas, trees for shade, and water features. Small lawn areas for informal seating and sunbathing were also well used. Visibility of the park from nearby streets was important to create a sense of safety. Cafés and other opportunities to purchase food also generated more activity and park use. Whyte's insights have helped in the creation and redesign of many urban parks, including the award-winning renovation of Bryant Park in New York City. Whyte, like the majority of park researchers, equates "successful parks" with the number of users; the more crowded a park, the more successful. (See Jerold S. Kayden's essay in this volume.)

Although park use is one measure of success, simply observing the number of people in a park does not reveal what the public enjoys about either the park or what meaning it has for them. Some researchers have proposed that people develop an attachment for places, an emotional bond between themselves and a particular place (Shumaker and Taylor 1983).

In Ann Arbor, Michigan, urban park users with a strong attachment to their nearby parks were eager to show them to other people and would experience some sense of loss if these parks were changed adversely. Moreover, park users who had a strong attachment for their nearby parks were more willing to become advocates for them in the political arena (Ryan 1997, 2000, 2005). People's love for place is often an unspoken but powerful motivation for intervention in the planning arena. Whyte had several places dear to his heart. The rolling countryside of Chester County, Pennsylvania, where he was raised, was under siege from urban sprawl in the postwar years and became the inspiration for his book *The Last Landscape* (1968), a seminal treatise on open space planning and conservation. His love for the vibrant, chaotic, and inherently unplanned use of sidewalks in New York City inspired his work on urban streets and plazas.

Park planners and managers need to understand what factors might lead people to develop an attachment to urban parks. Whyte was correct in his intuition that park use is an important measure of success; it is also an important factor in

Figure 1   Bicyclists in Chicago's Lincoln Park. (Photo by Robert L. Ryan.)

creating an attachment between people and place. Attachment to urban parks may manifest itself in people having an emotional connection or strong affinity for the place itself, as well as feeling that a particular park is the best place to engage in recreation activities such as walking or biking (figure 1). A study of rail-trail users in three urban areas (Dubuque, Tallahassee, and San Francisco) found that those who were frequent trail users and who lived closer to the trails expressed a stronger attachment than those who did not (Moore and Graefe 1994). As Whyte also ascertained, the physical features of an urban park have a profound effect on whether people will use the park or not; thus, the physical attributes of a park may also be key factors in creating an attachment between people and place. Research has found that people may develop strong attachments to certain trees or woods (Dwyer, Schroeder, and Gobster 1994). It is important to know what other physical features contribute to developing an attachment for urban parks.

Research on What Causes Attachment to Urban Parks

A study of three urban parks and natural areas in Ann Arbor, Michigan, provides insights into what factors may contribute to people's attachment for urban parks (Ryan 1997, 2000, 2005). In particular, this study focused on the influence of park use (i.e., experience) and the place itself (i.e., the physical attributes of the place) on the public's attachment to these parks. An important contribution of this study was to expand the traditional definition of park user, as used by Whyte and other recreation researchers, beyond simply those who are physically using a particular park to a broad range of people who have some type of experience with the park. The 328 participants in the Ann Arbor urban parks study included those involved in park design, those who maintain the parks (park staff and volunteers), recreation users (many of whom also lived near the parks), and those whose only use of the parks was visual enjoyment while passing by.

Using photos of individual parks, the survey sought to ascertain patterns and frequency of usage as well as the respondents' opinions on park design and management. The survey found that all these different types of users had an attachment to their nearby parks, including those who only viewed them from their home or car, without entering the park. This last category in particular, park neighbors and passersby, is most often missed in park studies and research. The type of experience people had in the park had an effect on the strength of their attachment to it. Local people, especially those who lived near the parks, had a stronger attachment to these particular parks than did either the park staff or volunteers. The more frequently people used the park for walking, biking, and other types of recreation, the stronger their attachment was for that place. People who volunteered in the parks expressed an attachment to their volunteer sites. A subsequent study also found that park and natural area volunteers also expressed greater appreciation for local natural areas in general (Ryan, Kaplan, and Grese 2001). Thus, it appears that encouraging park use in many different forms helps foster an attachment between the public and their urban parks.

Physical features within parks also influenced the level of attachment to parks. Certain parks were more "loved" than others, as were certain places within each park. For example, a riverside university arboretum in one park elicited much higher attachment ratings locally than did a restored prairie area. Park staff and volunteers, however, also appreciated the more overgrown areas of the parks where native vegetation was being encouraged. The study found that the more that people knew about the benefits of native plants and ecosystems, the stronger their appreciation for native plantings versus ornamental plantings, a significant finding for park planners and managers seeking to enhance the biodiversity in urban parks. Volunteer programs and other educational outreach programs can help improve the public's acceptance of native landscaping. Other findings from that study (Ryan 2000, 2005), however, suggest that native plantings must be designed and managed in a manner that fits with the public's expectations. Strategies for incorporating native plants in a manner that is appreciated by the public are discussed later in this essay.

Various types of users viewed park management differently. Those who only viewed the parks from home or street preferred traditional management: neatly mown lawns and clipped shrubs. Park neighbors were also concerned that increased development in the parks, such as building additional parking areas or visitor centers, would bring more outsiders to the parks and increase traffic. Many of the more active users—walkers, bikers, and bird-watchers—wanted park managers to let nature take its course. Park volunteers and staff preferred management to promote native species, such as removing nonnative trees and shrubs and using controlled burns to enhance native grassland areas and woodland understory plants. These conflicting preferences tend to complicate park management. Park planning and management must involve diverse types of users and try to reconcile diverse needs.

Strategies for Nurturing Attachment to Urban Parks

Strategies for promoting a connection between the public and their parks include (1) understanding existing park features and uses, (2) improving visibility and perceptions of safety, (3) incorporating design features that promote park use, (4) providing opportunities for the public to adopt their parks as part of volunteer stewardship programs, and (5) making small-scale improvements. These strategies are discussed in turn.

Understanding Existing Park Features and Uses

Using the physician's motto "Do no harm," park planners, designers, and managers need to respect the attachment that people may have for existing parks or features within them, including those features not planned or officially promoted. As Whyte revealed in his research, there may exist urban parks and plazas that already function very well for their users in diverse ways. Particular specimen trees or other features may already be "sacred" places to local residents. Behavioral mapping, as employed by Whyte, helps us understand how an existing site is currently being used. This technique has been refined by the New York–based Project for Public Spaces, Inc. (2000a). For example, a vacant urban lot may be used by local bird-watchers who appreciate the variety of species that use successional vegetation. Interviews and surveys of park neighbors, users, volunteers, and staff, however, are also needed so that we can understand the deeper meaning that these places have for local people and can understand why some places are used, whereas others are not. Local people often have insights about a particular park that are difficult for professional park planners and managers to ascertain. Drawing on this local knowledge is the key to designing a park that does a better job meeting the open space needs of the community (Kaplan, Kaplan, and Ryan 1998; Project for Public Spaces 2000a).

Because attachment to a park is strongly associated with use of it, park planners and managers need to develop ways to encourage park use by diverse groups

(Whyte 1988; Kaplan, Kaplan, and Ryan 1998; Marcus and Francis, 1998), which requires an understanding of user needs. For example, groups that recreate in extended families need larger picnic areas and shelters. Other groups may require accessible nature trails and boardwalks.

Improving Visibility and Safety

Enhancing the visibility of parks from nearby homes and streets helps establish visual connection to the larger public realm. People who rarely venture into a park may nevertheless develop a sense of attachment to it. The beautiful maple tree outside one's office window or the park view from one's apartment offers respite from the harried urban world (figure 2). Furthermore, neighbors and park users are the self-appointed guardians of many urban parks, and they will protest negative changes to their parks, such as removing trees, paving over parkland, or intrusions by private commercial interests.

Conversely, park users may gain a sense of safety within the park if they can see nearby homes and streets. For example, before the renovation of Bryant Park in

Figure 2   Boston's Commonwealth Avenue Park. Homes within view of a park may enhance a sense of safety for park users. It is also important, however, to use canopy trees and other screening devices to provide park users with some sense of enclosure from surrounding urban land uses. (Photo by Robert L. Ryan.)

New York City, large hedges surrounding the park made it difficult to see who was using the park. The park was perceived as an unsafe place and became the haven for drug dealers and other criminal activity. One of the key changes to the park's successful renovation in the 1980s was removing these hedges and increasing visibility from nearby streets and office building (Project for Public Spaces 2000a).

Of course, promoting visibility between a park and its environs conflicts with affording privacy and refuge from the outside world. Providing a sense of enclosure is often a key element in creating a place that people enjoy (Kaplan, Kaplan, and Ryan 1998). In busy urban areas, a canopy of large trees can provide screening from nearby buildings, yet still allow visibility from nearby streets. The placement of low shrubs and fences is another strategy to delineate a park space while still allowing visual access. In areas where taller screening, such as walls or hedges, is needed to hide unsightly views or buildings, gateways or breaks in screening elements can provide visibility from key vantage points as well as act as entries to the park and orient new visitors (Kaplan, Kaplan, and Ryan 1998).

Park Design Features That Promote Use

Some design features that foster attachment to parks include the following:

- Providing a variety of seating options
- Creating comfortable microclimates
- Incorporating well-designed water features
- Responding to the needs of a diverse range of users
- Increasing park activity with food vendors and festivals
- Promoting volunteer stewardship activities.

Some of these design strategies were discussed earlier when reviewing Whyte's work. For example, providing a variety of seating options allows for different-sized groups to meet, including individuals, couples, and larger groups. Movable chairs are especially appreciated because they allow people to customize their own seating arrangements. Comfortable seating options are also important. For example, benches with backs and armrests are easier for the elderly and those with disabilities to use. Creating comfortable microclimates for seating within an urban park is essential to ameliorate the temperature extremes of the urban environment. Shelter from strong winds and sunny areas can extend park use in colder climates. Shade is important in warmer climates and during the summer months. Whyte (1988) also found that well-designed shady plazas were also heavily used in the cooler months of the year.

The importance of landscape features such as trees is important in urban park and plaza design. Vegetation, however, must be used in a manner that creates a preferred setting rather than an overgrown, densely planted, or chaotic design. The presence of water is an attraction in many parks, but like the use of vegetation, the

Figure 3 Allowing park vendors and food kiosks is one strategy to increase park activity and use as shown here along San Francisco's waterfront. (Photo by Robert L. Ryan.)

quality and design of water features can influence people's attraction to it. Whyte (1988) suggests that water features in urban plazas should allow people contact with the water, dangling their feet in fountains or even splashing around in them, yet many urban plazas discourage this type of activity. The quality of the water and the edge treatment are key to the public's appreciation for water features (Kaplan, Kaplan, and Ryan 1998). Eroded stream banks are not preferred by the public, neither are overgrown or polluted appearing water bodies, even if the algae bloom is natural in occurrence. Likewise, natural-edged water features with vegetation are generally more preferred than hard-edged water bodies.

In general, parks with more activity have increased usage, which in turn can increase perceptions of safety within the park. Food vendors and other kiosks attract the public to parks and can generate revenue for park maintenance (Project for Public Spaces 2000a, 2000b). Increased activity can also come from festivals and other seasonal activities such as farmers' markets and concerts in the park (figure 3). Designing a park for a diverse range of users can also increase park activity. For example, creating spaces that respond to the needs of children, teens, adults, and the elderly can foster park use at different times of the day. Providing for a range of park uses from active recreation (e.g., sports fields and playgrounds) to passive recreation (e.g., bird-watching, picnicking, and walking) can increase the diversity of park uses and potential park stewards. Single-purpose parks such as sports fields have a narrow clientele. Successful urban parks, such as Central Park, have found ways to incorporate sports fields in a manner that still allows other more informal uses.

Volunteer Stewardship Activities

In addition to encouraging traditional use of urban parks, another strategy for nurturing an attachment between people and parks is to create opportunities for the public to participate in park design and management. Volunteer stewardship programs have become a driving force in revitalizing urban parks in the United States (figure 4). There is preliminary research to suggest that continued participation in volunteer activities, particularly in environmental stewardship programs, promotes a sense of attachment and increased appreciation for urban natural areas (Ryan, Kaplan, and Grese 2001). Although additional research is needed to understand how other volunteer activities, such as flower plantings and urban gardening, promote an attachment and a sense of ownership by local residents, there is ample anecdotal evidence to suggest that getting the public involved in hands-on management and improvements to local parks and other urban open spaces creates the local stewards that are essential for the survival of urban parks (Project for Public Spaces 2000b). Tree-planting projects and other horticultural activities require an ongoing commitment by local volunteers to maintain and nurture these plantings. Watering, pruning, and weeding require that volunteers are frequently working in the parks, thus increasing the activity that helps make parks safe. The results of these labors—new trees and flowers where there were previously weeds or vacant lots—show the public that someone cares about these places. From volunteers' perspective, as the investment of time and energy increases, so might their attachment and sense of ownership for the particular park or garden in which they are volunteering.

Volunteers, both temporary and long term, are strongly motivated by the opportunity to learn new knowledge and skills (Grese et al. 2000; Ryan, Kaplan, and Grese 2001). Providing opportunities for volunteers to learn more about the cultural and natural history of the parks in which they are working can help encourage volunteer participation. Volunteer activities as well as educational programs can also help increase the public's appreciation and acceptance of environmental restoration efforts.

Research suggests, however, that the public's appreciation for native landscaping requires more than simply environmental education. Native plantings should exhibit a sense of intentional management. Landscape architect Joan Nassauer (1995) proposes that "cues to care" be used with native plantings to improve the public's acceptance for them. These cues to care include mowing the edges of native grass areas, pruning shrubs and trees, using more intensive native flower plantings, and

Figure 4 Volunteering in urban parks can increase the public's stewardship of urban parks while also providing many tangible park improvements. In this photograph, volunteers in Ann Arbor, Michigan, are helping to restore a natural area within an urban park by removing non-native invasive shrubs. (Photo by Robert L. Ryan.)

placing fences, birdfeeders, and other landscape elements in the park that signify human presence in the native landscape. Research in urban natural parks in California (Matsuoka 2002) and southeastern Michigan (Ryan 1997, 2000) has shown an increased appreciation and perception of safety where native plantings appear intentionally managed. Such management activities—native flower plantings and pruning—can also involve park volunteers in stewardship of native landscape plantings within urban parks.

Small-Scale Improvements

Finally, small-scale, incremental improvements help nurture an attachment between people and their parks. Because the public may already have an attachment to certain aspects of a park, small changes allow managers and planners to gauge the public's response before making changes that could be perceived as catastrophic by park users and neighbors (such as major tree cutting or burning of prairie areas). The idea of using park improvements as small experiments requires that managers are able to track the effect of design and management changes on the public's use of a park, perceptions of safety, aesthetic appreciation, and other variables that may be important (Kaplan, Kaplan, and Ryan 1998). By understanding the positive and negative effect of park improvements, park managers and designers are better able to practice adaptive management that can respond to the changing context of urban parks, a necessity because park users, neighborhoods, and resources are often in a state of flux.

Small-scale changes have an additional benefit of showing immediate, tangible results to urban residents who have often been waiting quite a long time to see some improvements in the face of urban park decline and neglect. Urban greening projects undertaken by such groups as the Pennsylvania Horticultural Society's Philadelphia Green program, the Horticultural Society of New York, and the Greening of Detroit have transformed many vacant lots, alleys and other public spaces into valuable community gardens and parks using this principle. (See the respective websites of these organizations for more information: www.pennsylvaniahorticul turalsociety.org/pg, www.hsny.org, and www.greeningofdetroit.com.) The added benefit of many of these urban greening projects is that they have also engaged the community in creating these spaces. Research has shown that volunteers in environmental stewardship programs are motivated by the ability to see some tangible results to the environment that result from their efforts (Grese et al. 2000; Ryan, Kaplan, and Grese 2001). Urban greening projects also show visible results of civic improvement. For example, the Horticultural Society of New York's Read and Seed program creates children's gardens in front of public libraries for inner-city children to enjoy as part of their summer reading program (Smith 2002). These small-scale improvements to urban parks and other public spaces can create tremendous positive effects with small amounts of time and energy.

The public develops strong attachments for many urban parks and natural areas. Affection for parks is a powerful stimulus to preserving, sustaining, and restoring urban parks and conservation areas. Park designers and managers, however, can only tap into this force if they incorporate multiple viewpoints into planning and management decisions. The public perceives urban parks through different lenses: as a green view outside one's window, a beautiful park on the drive to work, a place to plant and nurture, or a place to recreate and relax. There is a strong need to expand the definition of park users beyond conventional ones such as dog walkers, children in playgrounds, and parents with strollers. Only then can the wealth of experience, as well as opinions, about how these parks should be improved and managed be captured. Sustaining urban parks requires increasing opportunities for the public to volunteer in maintaining, expanding, and improving these valuable resources. At the same time, encouraging park volunteers opens another avenue for people to develop ownership and attachment for their nearby parks.

Nurturing public spaces is a job that is never finished (Project for Public Spaces 2000a). Rather than seeing this task as a negative, it can be seen as providing myriad opportunities for engaging the public with their parks. Creating and sustaining urban parks provide a lifetime of challenges for those who love the precious urban green spaces that define a humane metropolis.

Acknowledgments

Thanks to the USDA Forest Service, North Central Forest Experiment Station, Evanston, Illinois, for its assistance in funding my research work on urban parks through Co-operative Agreement 23-96-06. Additional thanks goes to my doctoral dissertation committee—Rachel Kaplan, Donna L. Erickson, Raymond DeYoung, and Stephen Kaplan—for their invaluable advice on the Ann Arbor parks study.

References

Benepe, A. 2002. Presentation at the Humane Metropolis Conference, People and Nature in the 21st Century City: A Symposium to Celebrate and Continue the Work of William H. Whyte. New York University, New York. 6–7 June.

Dwyer, J. F., H. W. Schroeder, and P. H. Gobster. 1994. The deep significance of urban trees and forests. In *The ecological city: Preserving and restoring urban biodiversity,* ed. R. H. Platt, R. A. Rowntree, and P. C. Muick. Amherst: University of Massachusetts Press.

Grese, R., R. Kaplan, R. L. Ryan, and J. Buxton. 2000. Psychological benefits of volunteering in stewardship programs. In *Restoring nature: Perspectives from the social sciences and humanities,* ed. P. H. Gobster and R. B. Hull. Washington, DC: Island Press.

Harknik, P. 2000. *Inside city parks.* Washington, DC: Urban Land Institute.

Hough, M. 1994. Design with nature: An overview of some issues. In *The ecological city: Preserving and restoring urban biodiversity,* ed. R. H. Platt, R. A. Rowntree, and P. C. Muick. Amherst: University of Massachusetts Press.

Kaplan, R., S. Kaplan, and R. L. Ryan. 1998. *With people in mind: Design and management of everyday nature.* Washington, DC: Island Press.

Lutz, D. 2002. The cuts go deeper. *Urban Outdoors* 88 (10 December). Newsletter of the Neighborhood Open Space Coalition and Friends of Gateway, www.treebranch.com.

Marcus, C. C., and C. Francis. 1998. *People places: Design guidelines for urban open space.* 2nd ed. New York: Wiley.

Matsuoka, R. 2002. Increasing the acceptability of urban nature through effective cues to care: A study of Lower Arroyo Seco Natural Park, Pasadena, California. Master's thesis. California State Polytechnic Univ., Pomona.

Moore, R. L., and A. R. Graefe. 1994. Attachment to recreation settings: The case of rail-trail users. *Leisure Sciences* 16:17–31.

Nassauer, J. I. 1995. Messy ecosystems, orderly frames. *Landscape Journal* 14(2): 161–70.

Platt, R. H., R. A. Rowntree, and P. C. Muick, eds. 1994. *The ecological city: Preserving and restoring urban biodiversity.* Amherst: University of Massachusetts Press.

Project for Public Spaces. 2000a. *How to turn a place around: A handbook for creating successful public spaces.* New York: Project for Public Spaces.

Project for Public Spaces. 2000b. *Public parks, private partnerships.* New York: Project for Public Spaces.

Ryan, R. L. 1997. *Attachment to urban natural areas: Effects of environmental experience.* PhD diss. University of Michigan, Ann Arbor.

————. 2000. Attachment to urban natural areas: A people-centered approach to designing and managing restoration projects. In *Restoring nature: Perspectives from the social sciences and humanities,* ed. P. H. Gobster and R. B. Hull. Washington, DC: Island Press.

————. 2005. Exploring the effects of environmental experience on attachment to urban natural areas. *Environment and Behavior* 37:3–42.

Ryan, R. L., R. Kaplan, and R. E. Grese. 2001. Predicting volunteer commitment in environmental stewardship programmes. *Journal of Environmental Planning and Management* 44(5): 629–48.

Shumaker, S. A., and R. B. Taylor. 1983. Toward a clarification of people-place relationships: A model of attachment to place. In *Environmental psychology: Directions and perspectives,* ed. N. R. Feimar and E. S. Geller. New York: Praeger.

Smith, A. R. 2002. *Greenways: The community outreach programs of the Horticultural Society of New York.* Memorandum, June.

Spirn, A. W. 1984. *The granite garden: Urban nature and human design.* New York: Basic Books.

Trust for Public Land. 2002. Americans vote to protect land. *Land and People* 14(1): 4. Also available as LandVote 2001 report by the Trust for Public Land and the Land Trust Alliance, www.tpl.org and www.lta.org.

Whyte, W. H. 1968. *The last landscape.* New York: Doubleday. Republished Philadelphia: University of Pennsylvania Press, 2002.

————. 1980. *The social life of small urban spaces.* New York: Project for Public Places.

————. 1988. *City: Rediscovering the center.* New York: Doubleday.

Respecting Nature's Design
in Metropolitan Portland, Oregon

Michael C. Houck

> *Instead of laying down an arbitrary design for a region, it might be in order to find a plan that nature has already laid down.*
> WILLIAM H. WHYTE, *The Last Landscape*

> *The belief that the city is an entity apart from nature and even antithetical to it has dominated the way in which the city is perceived and continues to affect how it is built. The city must be recognized as part of nature and designed accordingly.* ANNE WHISTIN SPIRN, *The Granite Garden*

Securing Urban Green Infrastructure

Henry David Thoreau's aphorism "In wildness is the preservation of the world" has driven the conservation agenda in the United States for over a century. The emphasis has been, first and foremost, the protection of wilderness, pristine habitats, and agricultural lands in the rural landscape. If we hope to succeed in protecting rural resource lands in the twenty-first century, a new corollary to Thoreau's mantra might be "In livable cities is preservation of the wild." We must commit significantly more attention and resources to the protection and restoration of natural resources in the urban landscape as a strategy for protecting farm, forest, and other rural resource lands. By creating livable urban communities, we will build public support for a smart growth agenda. Through higher density, compact cities will promote enhanced protection of the rural landscape from urban sprawl. The quid pro quo, however, must be the protection and, where necessary, restoration of a vibrant urban green infrastructure of healthy streams, fish and wildlife habitat, parks, and recreational trails where the vast majority of our population lives: namely, in our cities.

In 1982, when I began my work as Audubon Society of Portland's urban naturalist, local planners believed that Oregon's land use program did not contemplate protection of natural resources inside our Urban Growth Boundary (UGB). The UGB, they believed, was to halt urban sprawl and to protect farmland and forestland outside the city. In fact, the argument has been made that protecting fish and wildlife habitat and too much open space inside the UGB was antithetical to good urban planning. Accordingly, the Portland metropolitan region has more than three hundred miles of streams that have been placed in underground conduits, and more than two hundred miles of streams and rivers are "water quality limited" or polluted, according to the state's Department of Environmental Quality. The steelhead trout and chinook salmon are listed as threatened under the federal Endangered Species Act, and the cutthroat trout is likely to be listed soon.

Developing a Regional Parks and Greenspaces System

In 1989, the lack of natural resource protection, park deficiencies, and incomplete trail systems stimulated establishment of a "Cooperative Regional System of Natural Areas, Open Space, Trails, and Greenways, for Wildlife and People" in the Portland-Vancouver metropolitan region. This initiative built on earlier efforts of many regionalists, including the Olmsted brothers (John Charles and Frederick Law Jr.), Lewis Mumford, and the Columbia Region Association of Governments (CRAG), the predecessor to Metro. John Charles Olmsted, in *Report of the Park Board, Portland, Oregon, 1903,* wrote: "While there are many things which contribute to the beauty of a great city, unquestionably one of the greatest is a comprehensive park [system]" (Olmsted 1903, 14). He urged the integration of natural areas in a comprehensive park system that would "afford the quiet contemplation of natural scenery [with] rougher, wilder and less artificially improved [parks]." He also presaged interest in urban waterway and watershed management by noting: "Marked economy may also be effected by laying out parks, while land is cheap, so as to embrace streams that carry at times more water than can be taken care of by drain pipes. Thus, brooks or little rivers which would otherwise be put in large underground conduits at enormous public expense, may be attractive parkways" (Olmsted 1903, 20).

These themes were echoed in the 1971 CRAG regional open space plan: "For many persons in the city, the presence of nature is the harmonizing thread in an environment otherwise of man's own making. . . . Comprehensive planning should identify floodplains, wetlands, scenic, wildlife and recreational [areas]. Development should be controlled." The report also called for bistate cooperation between Oregon and Washington, a concern earlier expressed by Mumford in a 1938 speech to the City Club of Portland. According to CRAG, "It is yet to be seen whether the Portland-Vancouver urban community and the states can muster the drive, inspiration, the legal tools to develop a regional park and open space program." The jury is still out on this question, although significant progress has been made in the past decade.

Most significantly, the CRAG report for the first time called for the integration of Olmsted's comprehensive and connected park system with Mumford's regional approach to establish a regional open space program that would "relieve the monotonous and the mechanical by preserving and enhancing those environmental features that have already stamped the region with their unique form and character, which make it a very special place to live, the rivers, streams, and flood plains; high points that overlook the cityscape" (CRAG 1971, 3).

As recently as 1988, with a few notable exceptions, the Portland region had implemented few of the recommendations set forth by the Olmsteds, Mumford, or the CRAG report. What changed that dynamic was the merging of interests of park and natural area advocates with regional trail advocates. In 1984, the Audubon Society of Portland advocated for the establishment of a metropolitan wildlife refuge system. The timing was propitious, given that Metro had just initiated a regional park resources inventory. Portland Audubon Society, the 40-Mile Loop Land Trust, and other park and greenspace advocacy groups successfully argued that the region needed a new, regional perspective in natural resource protection and management.

The ideal agency to provide such a regional perspective was Metro, the only directly elected regional government in the United States. All twenty-five cities and three counties within its jurisdiction must, by law, amend their comprehensive plans to conform to regional regulations, developed through painstaking consensus-building among stakeholders.

An important first step toward a regional natural areas system was the production of a four-county, bistate natural areas map through collaboration of Audubon, Metro, and Portland State University's Geography Department. The map covered 364 square miles on the Oregon side of the Columbia River and 145 square miles in Clark County, Washington. As of 1990, 29 percent or 108,000 acres, of the region remained undeveloped, and of that total, only 8.5 percent was publicly owned. Nearly half of that was in Portland's 5,000-acre Forest Park. The knowledge of the scarcity of publicly owned land, combined with the prospect of more than a million new metropolitan residents by 2040, generated widespread political and popular support for a regional greenspaces program.

Another step was the arrangement of site visits by forty Portland and Vancouver elected officials, park professionals, and park and greenspace advocates to the California East Bay Regional Park District. The East Bay District, which serves Contra Costa and Alameda Counties (Oakland/Berkeley area), had recently passed a $225 million bond measure; meetings with their staff stimulated interest in similar efforts for Portland.

Public support was also generated by several "Country in the City" symposia held at Portland State University. Experts in regional and greenspace planning such as Dr. David Goode, then director of the London (U.K.) Ecology Unit, Tony Hiss, author of *The Experience of Place,* and Charles E. Little, author of *Greenways for America,* spoke at these events. The result was a groundswell of support from urban stormwater management agencies, park providers, and land use advocates to develop a more comprehensive approach to natural resource management in the Portland-Vancouver metropolitan region.

With support from Senator Mark O. Hatfield, then chair of the U.S. Senate Appropriations Committee, and Congressman Les Aucoin, Congress in 1991 appropriated $1,134,000 for the greenspaces program. The regional office of the U.S. Fish and Wildlife Service (FWS) administered the funds, and FWS field staff were assigned to work with Metro to ensure that the nascent greenspaces program remained true to its ecological focus. Other partners included the National Marine Fisheries Service, the U.S. Environmental Protection Agency, and the Oregon Department of Fish and Wildlife.

In 1992, the Metro Council adopted the *Metropolitan Greenspaces Master Plan,* which had the following goals (Metro Council 1992, 1):

1. Create a cooperative regional system of natural areas, open space, trails, and greenways for wildlife and people in the four-county metropolitan area.
2. Protect and manage significant natural areas through a partnership with governments, nonprofit organizations, land trusts, interested businesses and citizens, and Metro.
3. Preserve the diversity of plant and animal life in the urban environment using watersheds as the basis for ecological planning.
4. Establish a system of interconnected trails, greenways, and wildlife corridors.
5. Restore green and open spaces in neighborhoods where natural areas are all but eliminated.[1]

A 1992 bond measure failed by an 8 percent margin, owing primarily to a lack of campaign funding and political commitment. A second levy was approved by more than 63 percent of the region's voters in May 1995. This levy produced $135.6 million, 75 percent going to Metro for regional parks and the rest to local park systems, although both the regional and local shares were to be spent exclusively on natural area acquisition and trails. Metro's land acquisitions included fourteen regional "target areas" and six trail and greenway project areas. As of April 2005, more than 8,200 acres of land had been purchased, donated, or protected with conservation easements, well exceeding the original target of 6,000 acres. A second bond for $220 million is planned for the fall of 2006.

Local Park Initiatives, Portland Parks, and Recreation

Significant progress has been made at the local level as well. In the same period that a regional greenspaces initiative was being launched through Metro, much was changing in the City of Portland's Parks and Recreation Bureau. Oaks Bottom Wildlife Refuge, a 160-acre wetland in the Willamette River floodplain in the heart of downtown Portland, was designated as Portland Park's first official urban wildlife refuge (figures 1 and 2). Portland has since added 902 acres to its natural areas program, and the Portland City Council added $300,000 for natural area maintenance in 2001.

In 2001, a new Portland Parks Vision 2020 Plan stated:

The city's parks, natural areas and recreation programs are among the essential elements that create a livable, dynamic and economically vibrant city; . . . Linking parks with greenways, trails and paths provides greater recreational benefit; Portland Parks will promote regional strategies to protect natural resource values of wildlife corridors, including: integrating trail planning with Metro Title 3 Water Quality and Goal 5 Protection programs; [and] recreation planning with Portland's "River Renaissance" and with Portland's "River Recreation" Plans. (Portland Parks and Recreation 2001, 30)

Thus, for the first time since the 1903 Olmsted master plan, natural resources and natural resource management were seen as equal with and complementary to recreational facilities and neighborhood and community parks.

Regional Growth Management

Oregon's land use planning program has been extremely successful at containing urban sprawl. Between 1990 and 2000, Portland's metropolitan population expanded by 31 percent, while urbanized land increased by only 3 percent. By contrast, Chicago's regional population grew by 4 percent between 1970 and 1990 but its urbanized land area increased by 46 percent. Kansas City's population grew by 29 percent during the same period, and its land consumption was 110 percent.

The primary objectives of Oregon's planning program have been to protect prime farmland and forestland outside the UGB and to reduce infrastructure costs through compact urban form. The challenge is not whether to hold a tight urban growth boundary to protect these lands, but how to simultaneously maintain quality of life *inside* the UGB. Unfortunately, the manner in which local jurisdictions have applied the state planning goals has led to an inequitable distribution of parkland, loss of natural resources, degraded water quality, and disappearance of fish and wildlife habitat throughout the region.

The failure of most local governments to protect urban natural resources is corroborated by Oregon's *State of the Environment Report* (State of Oregon 2000, 108):

The annual rate of conversion of forest and farmlands to residential and urban uses has declined dramatically since comprehensive planning land use planning was implemented during the 1980s. *However, these laws were not written to address ecological issues, such as clean water or ecosystem function within urban growth boundaries. In order to meet the economic and social needs of humans, native vegetation and habitats may be destroyed and converted to buildings and paved surfaces.* (emphasis added)

Although the report's conclusion is debatable on both technical and legal grounds, it is functionally correct. The problem has not been the state land use planning program, but rather that local jurisdictions have implemented the program in a manner that has virtually ignored urban natural resource protection. The plan-

Figure 1 Discovering the wonders of urban nature. (Photo by M. C. Houck.)

Figure 2 Oaks Bottom Slough, Portland, Oregon's first designated urban wildlife refuge. (Photo by M. C. Houck.)

ning program, if implemented in a manner that seeks to protect natural resource values, can be used effectively. Metro, the City of Portland (through its Healthy Portland Streams efforts), and the City of Wilsonville have shown that if the political will, sufficient resources, and staff expertise are there, planning works to protect natural resources.

In 1994, the Region 2040 growth management planning process challenged the region to develop

> an integrated, multiobjective floodplain management strategy . . . which recognizes the multiple values of stream and river corridors including: enhanced water quality, fish and wildlife habitat, open space, increased property values, education, flood reduction, aesthetics, and recreation. An interconnected system of streams, rivers, and wetlands that are managed on an ecosystem basis and restoration of currently degraded streams and wetlands are important elements of this ecosystem approach. (Metro Council 1997, 9)

Metro took a page from Ian McHarg's *Design with Nature* (1968) in declaring more than 16,000 acres as "unbuildable," including wetlands, floodplains, two-hundred-foot buffers bordering streams, and slopes exceeding 25 percent. This action was consistent with McHarg's approach to subtract sensitive lands from the regional plan before determining the region's "carrying capacity" for homes, roads, and other infrastructure. Thus, Metro did not include the "unbuildable lands" when it calculated the acreage inside the UGB necessary to meet the region's development needs; such areas were simply placed out of consideration for future development. In 1996, Metro in its landmark greenspaces resolution called for expansion of the UGB if necessary to accommodate growth rather than sacrificing unbuildable lands within the UGB.

In 1998, Metro sought to protect the 16,000 acres of "unbuildable" lands by adopting "Title 3 of the Urban Growth Management Functional Plan," which affords *minimal* floodplain protection by requiring "balanced cut and fill" in floodplains if they are allowed to be developed. Title 3 also requires that fifteen-foot to two-hundred-foot vegetated corridors be protected along streams for water quality purposes. These regulations were challenged unsuccessfully at the Oregon Land Use Board of Appeals by Washington County, the cities of Tualatin and Tigard, homebuilder and real estate associations, and others.

Coalition for a Livable Future

Coalition for a Livable Future (CLF) was formed in 1994 by the Portland Audubon Society, 1000 Friends of Oregon, Community Development Network, Bicycle Transportation Alliance, Urban League of Portland, Ecumenical Ministries of Oregon, and others. Myron Orfield, a state legislator from the Minneapolis–St. Paul region, helped catalyze the formation of the coalition in 1994. He argued, based on his research on urban decay around the United States, that urban sprawl leads to the "hollowing out" of core cities, leaving behind pockets of poverty. Orfield found similar, albeit less extreme, trends appearing in the Portland metropolitan region.

CLF currently has more than sixty nonprofit organizations working in the Portland-Vancouver metropolitan region. They include a core group (1000 Friends of Oregon, Citizens for Sensible Transportation, Audubon Society of Portland, Willamette Pedestrian Coalition, Urban League, Bicycle Transportation Alliance, and the Community Development Network) that has been joined by suburban affordable housing representatives, stream and watershed groups, neighborhood associations, food policy advocates, and mainstream conservation groups like the Sierra Club and Oregon Environmental Council. CLF's stated mission is "to protect, restore, and maintain healthy, equitable, and sustainable communities, both human and natural, for the benefit of present and future residents of the greater metropolitan region." The focus of the coalition is to influence public land use, transportation, housing, economic, and environmental policies through advocacy, research, and public education. It has working groups on natural resources, food policy, transportation reform, urban design, religious outreach, economic vitality, and affordable housing.

Affordable Housing

In advocating affordable housing, CLF seeks to refute the contention that housing costs are a function of a tight Urban Growth Boundary, which is routinely blamed by the homebuilders for driving up the cost of land and housing, and regional land use planning. CLF's position is supported by a report from the Brookings Institution (2002) citing dozens of academic studies that undermine the contention that housing price increases in the Portland region have outstripped the national average. Mary Kyle McCurdy, urban development specialist for 1000 Friends of Oregon, states that "this report demolishes the tired argument that urban growth boundaries are to blame for a supposed crisis in housing affordability."

Public Transit

The coalition also works on transit issues. According to Tri-Met, the Portland regional transportation agency, there has been $2.9 billion of transit-oriented development—everything from apartments, mixed-use high-density developments to office buildings— along the existing east and west side rail lines since the opening of east side MAX (light rail line). Examples are Orenco Station in the city of Hillsboro to the west of Portland and infill mixed-income apartments on the east side. Tri-Met's average daily boardings are just under 300,000 a day, with MAX totaling more than 70,000 riders a day. MAX ridership has tripled in its fifteen-year history. Each weekday, MAX eliminates 48,000 car trips from the greater Portland roads; its ridership is increasing at about 5 percent a year.

Congressman Earl Blumenauer, who represents the Portland area, stated in his address "Portland: Ground Zero in the Livable Communities Debate" (Blumenauer 2000, 3): "Transit usage has increased 143 percent faster than the growth in population, and most critical, it has increased 31 percent faster than growth in vehicle miles traveled since 1990. For seven consecutive years, every month has shown an increase in transit ridership over the previous year. No other region can make that claim."

Farmland Protection

According to 1000 Friend's Robert Liberty, "The Oregon planning program (by) mandating urban growth boundaries around every city in the state [has protected] 40,000 square miles for farming, ranching and forestry (Accomplishments of the Oregon and Metro Portland Planning Programs 1998). Blumenauer concurs:

> From the top of a tall building in downtown Portland, you can see Sauvie's Island, prime farmland, a 10-minute drive to downtown Portland, flat and buildable. There is virtually the same amount of land in agriculture now as 25 years ago, which but for our land use planning laws, would have all been lost. In Washington County next to Portland, despite the addition of 40,000 people between 1982 and 1992, annual farm income increased 57 percent. At the same time, neighboring Clark County in Washington State lost 6,000 acres and farm incomes rose only 2 percent per year. Metropolitan Portland is the largest agricultural producing metropolitan area in Oregon.
>
> Some of our claims and accomplishments are overblown and we have fallen short of the mark in some areas. The mythic UGB is a prime example. It is not as powerful as it could have been. We placed too much emphasis on simply protecting farm and forest land, rather than creating livable communities. (Blumenauer 2000, 2)

Urban Design: Building It Greener, Lighter, Cheaper, and Smarter

Landscape architect Patrick Condon of the University of British Columbia has introduced Portland to his concept of making cities *greener, lighter, cheaper, and smarter,* as opposed to gray, heavy, expensive, and dumb, the manner in which most of our cities are built. His concept involves promoting high-density urban development while simultaneously reducing the effects of imperviousness on urban waterways, integrating the green infrastructure with the built environment. It is the final stage of achieving livable, compact urban form. We can protect all the streams in the metropolitan region, we can establish fish and wildlife management areas, and we can restore degraded aquatic resources. All that work, however, will have been for naught if we do not address the severe hydrologic impacts of urban stormwater on urban aquatic systems.

Effective imperviousness should be reduced by at least 10 percent, the threshold at which streams begin to "fall apart" owing to stormwater runoff, according to the

Center for Watershed Protection in Maryland. Compared with other regions of the United States, Portland's rainfall comes in smaller amounts spread out over longer periods. More than four-fifths of Portland's rainfall events are less than 0.5 inch, which affords the opportunity to capture and infiltrate much, if not all, of the water that would otherwise be conveyed by sewer pipes to the nearest stream.

Approximately 40 percent of all stormwater runoff in the Portland metropolitan region comes from transportation facilities. To address that issue, Metro has developed a design manual, *Green Streets, Innovative Solutions for Stormwater and Stream Crossings,* that focuses on design solutions that aim to reduce these stormwater effects on streams as well as the physical effects that road projects have on the riparian ecosystem. Elements of green streets include a system of stormwater treatment within rights of way, reduction of the volume of water piped directly to streams and rivers, incorporating the stormwater system into the aesthetics of the community, and minimizing effects of streets on streams or wetlands.

Furthermore, the City of Portland is rewriting its stormwater codes to reduce the effects of future development on aquatic systems and to promote the use of green roofs and ecoroofs in urban stormwater management.

Damascus Area Community Design Workshop

In late 2002, the Metro Council decided to expand the Metro urban growth boundary by more than 12,000 acres in response to state-mandated planning regulations that require a twenty-year land supply for housing and other development needs be provided inside the region's UGB. As areas near the UGB in Multnomah and Washington counties consist largely of either high-quality farmland or land that is topographically inappropriate for urbanization, Metro looked first to partially developed areas as Damascus in rural Clackamas County, in the southeast quadrant of the region when considering where to expand.

The Damascus area is an unincorporated community of approximately five thousand people located about twelve miles southeast of downtown Portland. The area is characterized by large-lot rural residential lands, small-scale nurseries, forested buttes, and significant fish and wildlife habitat. Transportation access to the rest of the region is poor. The Damascus Area Community Design Workshop was a community-based effort by the Coalition for a Livable Future to create a regional model for livable, equitable, and environmentally sound urban development in this possible UGB expansion area. The workshop applied design principles for urbanization that use land efficiently, protect and restore fish and wildlife habitat areas, protect natural stream flow, provide for a fair share of the region's new jobs, and include ample housing and transportation choices in every neighborhood. The workshop broadened the range of choices to be considered in designing newly urbanized areas, informed decisions to be made by Metro and Clackamas County officials as they consider a UGB expansion, and provided a model that can be adapted by other community design efforts in Oregon and other states. Recently, Damascus citizens voted to incorporate into the region's twenty-fifth city, and a planning process is under way that integrates most of the Coalition's Damascus Design Workshop recommendations into their vision for what they want their community to look like in the future.

When considered cumulatively, all these efforts hold great promise to create a just and sustainable metropolitan region. Passage of a property rights–fueled measure (Measure 37), which would require compensation or waiver of environmental regulations, in the fall of 2004 casts a long shadow over such efforts, however. As we go to press, Oregon's Marion County District Court has ruled Measure 37 unconstitutional. Creative local and regional planners and legal challenges will, I believe, prevail. The result will be a metropolitan region worthy of the lofty visions that John Charles Olmsted, William H. Whyte, Lewis Mumford, and Anne Whistin Spirn have envisioned.

Note

1. In March 2005, the newly appointed Greenspaces Policy Advisory Committee adopted a new vision for the creation of a comprehensive regional, bistate parks, trails, and greenspaces system. This document goes far beyond the original 1992 Greenspaces Master Plan in calling explicitly for a regional, bistate Biodiversity Protection and Management Plan and urging that the Portland region's work be linked to similar efforts in urban communities throughout the Willamette River Valley. In March 2006, Metro Council referred a $220 million bond to the region's voters for a November 2006 vote. This bond will allow the purchase of an additional 5,400 acres of natural areas, inside and outside the region's UGB.

References

Blumenauer, E. 2000. Portland: Ground zero in the livable communities debate. Address to CNU 2000: The Politics of Place. Eighth annual Congress for the New Urbanism, Portland, Oregon, 15–18 June.

Brookings Institution. 2002. *The link between growth management and housing availability: The academic evidence.* Available online at www.brookings.edu/urban.

CRAG [Columbia Region Association of Governments]. 1971. A proposed urban-wide park and open space system. 1 March.

McHarg, I. 1968. *Design with nature.* New York: Garden City Press.

Metro Council. 1992. *Metropolitan Greenspaces Master Plan.* July.

———. 1997. *Regional Framework Plan.* Chapter 4, Water management. 11 December.

Olmsted, J. C. 1903. *Report of the Park Board, Portland, Oregon.* With the Report of Messrs. Olmsted Bros. Landscape Architects, outlining a system of parkways, boulevards, and parks for the City of Portland.

Portland Parks and Recreation. 2001. *Parks Vision 2020 Plan.* July.

Spirn, A. W. 1985. *The granite garden: Urban nature and human design.* New York: Basic Books.

State of Oregon. 2000. *State of the Environment Report.*

Whyte, William H. 1968. *The last landscape.* New York: Doubleday. Republished, Philadelphia: University of Pennsylvania Press, 2002.

Promoting Health and Fitness through Urban Design

Anne C. Lusk

Sixty-five percent of the U.S. population is now overweight, and the resulting negative health consequences include premature death, cancer, heart disease, diabetes, stroke, and other chronic diseases (U.S. Department of Health and Human Services 1996, 2000). This rise in obesity is a result of poor diet and physical inactivity or an energy imbalance from an increase in caloric intake and a decrease in physical activity. In 2002, 25 percent of Americans did not participate in any physical activity during the preceding month (Centers for Disease Control and Prevention 2002), and in 2003, 38 percent of students in ninth through twelfth grades viewed three or more hours of television a day (Centers for Disease Control and Prevention 2004). Physical activity provides a variety of physiological and psychological health benefits; therefore, recommendations were made for thirty to ninety minutes of moderate physical activity most days of the week (Pate et al. 1995; Department of Health and Human Services and Department of Agriculture 2005). Interventions, such as the provision of facilities or the creation of programs, can be effective ways to combat obesity by increasing levels of physical activity (Kahn et al. 2002). Certain changes to urban forms could enable the physical activity as a routine part of the day (Handy et al. 2002; Sallis, Kraft, and Linton 2002; Killingsworth et al. 2003). A critical element of a "humane metropolis" is therefore to alleviate personal discomfort, depression, and poor health through encouraging outdoor physical activity and exercise within the urban environment. This essay is concerned with what urban design features would be the most "humane" and encourage more people to engage in physical activity.

The Nineteenth-Century Sanitary Reform Movement

The building of dwellings to accommodate the astronomic increase in urban populations in the industrializing nations during the nineteenth century lagged far behind demand. Overcrowding to inhuman levels was ensured by the prevailing building practices of the times. Unfettered by any public regulations, tenement building was a joint result of (1) the need to be within walking distance to employment and to family members and (2) the builder's greed for profit. Thus, dwellings were minute in size and packed together, with space left unbuilt only to the minimum extent necessary to provide physical access to each unit (Platt 2004, 99). One result of this pervasive overcrowding and lack of fresh water, daylight, and drainage was a series of epidemics that ravaged most European and North American cities during the early nineteenth century, including New York City.

Beginning in the 1830s, the progressive reformer Edwin Chadwick conducted the first studies of sanitary conditions in the industrial slums of England. Through crude geographical surveys and the new science of statistics, he related the spatial incidence of infectious disease to overcrowding and sanitary deficiencies. A series of reports that he prepared for the Parliamentary Poor Laws Commission laid a basis for the eventual adoption of Great Britain's first public health act in 1848. Chadwick's work in turn inspired comparable investigations by public health reformers in other cities, notably including New York City. In 1845, John H. Griscom published a landmark report entitled *The Sanitary Condition of the Laboring Population of New York* that was directly modeled on Chadwick's work. Based on his own studies of slums in lower Manhattan, Griscom called for a wide range of improvements, including a public water supply, parks, and public control of building design and occupancy. The New York (State) Metropolitan Health Act of 1866 was the first major U.S. law in this field (Platt 2004, 105). (The U.S. federal government would play no major role in urban environmental issues until a century later.)

City Form, Neighborhoods, and Parks

In 1895, the New York State legislature appropriated $5 million to condemn certain tenements and create small parks in the crowded slums of Manhattan. The population at the time was not necessarily overweight, but people were in desperate need of more sanitary conditions than those available in a city that was 30 percent more crowded than Prague, Europe's least livable city (Scott 1969). The social reform movement, through physical environmental determinism, thus sought to improve the health conditions of the poorest residents by providing some greenspaces and playgrounds amid the tenements of Five Points and its environs described by Jacob Riis as "the wickedest of American slums" and the "foul core of New York's slums" (quoted in Page 1999, 73). The demolition of buildings and creation of parks introduced health-inducing sunshine, fresh air, and open space to the immigrant population.

New York's Central Park, designed in the 1850s by Frederick Law Olmsted and Calvert Vaux, is said to have been fashioned after Birkenhead Park in England and designed for the masses who had no access to the healthy outdoors (Rybczynski 1999). Olmsted wrote about the need for urban inhabitants to escape unhealthy urban congestion, and, if they did not possess funds to leave the city, they should have places for respite and relaxation within reach of their homes (Olmsted 1865).

Central Park at the time, however, was still distant from the "huddled masses" of the Lower East Side, requiring the poor either to pay for a horse-drawn omnibus or walk an hour or more each way along muddy thoroughfares. Moreover, the paths in Central Park were designed for carriages, bridle riding, or walking. Because riding in a carriage was sedentary, the only real physical activity in Central Park would have been horseback riding or walking. The walks, especially by women with long skirts, might have been more apt to be taken passively on narrow paths or within the greenswards rather than on the many bridle paths that were dominated by horses. The park was not designed for bicyclists because the basic bicycle was not invented until the 1870s and reinvented in 1890 when pneumatic tires and a chain drive were added. Thus, Central Park, until modified for more fitness-oriented uses in the later twentieth century, was largely a scenic amenity for the higher classes of the city with little contribution to the fitness of the general population.

Olmsted also designed residential developments with park components, most notably Riverside, a suburban community near Chicago built in 1869. Olmsted's plans for Riverside included sidewalks, macadam roads, provisions for street cleaning, convenient transportation, and access to bakeries and stores. The best land was to be set aside as public grounds that included playgrounds, commons, and village greens. Privacy and control of the yard would be bequeathed to each individual homeowner, but there would also be communal space for play and socializing (Olmsted 1868; Fisher 1986). These communal spaces were then available to the residents wealthy enough to own homes in the developments; they were not considered public parks for all the residents of Chicago.

The English progressive reformer Ebenezer Howard launched the garden city movement in Great Britain and the United States with his famous tract, *To-morrow: A Peaceful Path to Real Reform* (published in 1898 and reissued in 1902 as *Garden Cities of To-morrow*). According to Lewis Mumford in his preface to the 1965 republication, *"Garden Cities . . .* has done more than any other single book to guide the modern town-planning movement and to alter its objectives" (Mumford 1965, 29).

Based on a population size of 30,000 per community, garden cities were to have as dominant features parks, tree-lined avenues, and public gardens (Ward 1992). Such ideal towns were to incorporate manufacturing, retail, and outdoor exercise facilities. They were to be surrounded by "greenbelts" of agriculture and forestland, but connected to a large metropolis by train. Pure garden city models were rarely actualized, but the ideological imprint remained. Howard's theory influenced the design of two prototype garden cities, at Letchworth and Welwyn, both near London. Through the advocacy of his disciples, the garden city movement influenced the British postwar "New Town" program, but with very different results from the small Victorian suburbs of Howard's concept.

Howard's ideas influenced the development of a few garden cities in the United States, most notably Radburn, New Jersey, designed in the 1920s by Henry Wright and Clarence Stein. Radburn included homes with living rooms facing long, connected parks with a bicycle path. This path connected to the schools and playground areas and featured bridges so that path users did not have to cross traffic. The vast postwar growth of suburbs in the United States, however, were emphatically oriented toward the automobile, with little provision for outdoor exercise other than school fields and playgrounds.

In 1929, New York's master builder and power broker Robert Moses created Jones Beach as an oceanside park built within reach of Manhattan for the working middle class unable to afford vacation homes in the country. The park included bathhouses, restaurants, expanses of beach, and vast parking lots and was accessible by two landscaped parkways and a train. Although during the Depression few people had cars, access was intended to be largely for those arriving by private car. Moses's parkways had overpasses built too low to allow buses, suggesting he did not want the "teeming masses" from the inner city to be flocking to his new parks (Caro 1974).

At about the same time, planner Clarence Perry, who grew up in another of Olmsted's planned communities, Forest Hills, Illinois, wrote the landmark *Plan for New York and Its Environs* for the New York Regional Plan Association. Among its proposed neighborhood design principles, the plan called for limited community size, inclusion of local shops, and the establishment of small parks and recreation spaces (Perry 1929). The early neighborhoods that followed these principles included communities such as Levittown on Long Island, started in 1947 and Park Forest, started in 1948. Such planned suburbs allocated homes, shopping, schools, and recreation to separate districts, often isolated from one another. They did, however, provide sidewalks on both sides of all the streets for easy access to destinations, an element sadly omitted from many more recent subdivisions. Levittown included village centers with a few retail stores, but it was difficult to combine the tasks of shopping with recreation. The chain-link-fenced recreation fields were usually distant from the commercial districts.

Individuals in communities such as Levittown who benefited most from open spaces without accompanying store traffic were the adjacent property owners who had long expanses of maintained parkland for a year-round view. Owners of property adjacent to Central Park also have views of magnificent parkland without storefronts on their personal sidewalk street fronts. The combination of stores and parks is beneficial because it allows people to "trip chain," or combine a leisure trip to the park with a purposeful trip to the store. In 1887, Mulberry Bend Park in New York City achieved the goal of providing pleasure with purpose in a park bordered by shops as seen in photographs by Riis (Alland 1975). The Central Park and Levittown decisions to not combine parks and shopping were based on economic and not physical activity reasoning because residences that bordered parkland could demand a higher premium.

Travel Corridors and Destinations for Walking and Cycling

In the late 1960s and early 1970s, urban planners specializing in behavioral design started writing about the built environment and perceptions of users. Kevin Lynch, Donald Appleyard, and John Myer developed symbols in their book *The View from the Road* and were able to identify what Lynch characterized as the paths, edges, districts, nodes, and landmarks as part of through-travel landscape analysis viewed from a car (Appleyard, Lynch et al. 1966). Lynch, in his book *The Image of the City*, wrote of the benefits of designing a path that provided a "classical introduction-development-climax-conclusion sequence" (Lynch 1960, 99). In *The Last Landscape*, urbanist William H. Whyte (1968, 325) conjectured that "people take much longer walks if they can see the building they are heading to." This visible building could be considered a "landmark" in Lynch terminology.

A Pattern Language by Christopher Alexander and others (1977) offered a "language" for building and planning with preferred design elements that would improve quality of urban life: "People find it easier to take a walk if they have a destination. This destination may be real, like a coke shop or a café, or it may be partly imaginary, 'let's walk round the block.' But the promenade must provide people with a strong goal" (Alexander et al. 1977, 172). Whyte and Appleyard also focused on the design of what they characterized as livable streets. Rather than allow domination by vehicles, people on foot or riding a bicycle should be accommodated, and the environment should provide opportunities for socializing, and greenery (Moudon 1987). Much of the focus of both Appleyard and Whyte, though, was on the pedestrian.

In 1965, a local citizen of Davis, California, Frank Child, wrote a letter to the editor stating that Davis should provide a safe environment for bicyclists who were increasing in numbers owing to an ever-expanding college population. Resistance from Davis's Select Board prompted a petition for bicycle provisions that was signed by hundreds of residents; that petition encountered even more official resistance. Reelections brought in two new and sympathetic selectpersons and a variety of designs were tested. One, placing a protected bicycle lane between the sidewalk and parked cars based on the European model, was rejected because the road bicyclists felt unsafe at the intersections with cars making right-hand turns. Bob Sommers and Paul Dorn, both psychologists, obtained funding to gather data on the bicycle facilities. Learning of the successful designs in Davis, many individuals visited the community to learn from the experience (Lott 2003). In 1972, the Federal Highway Administration signed a contact with Deleuw-Cather in San Francisco to write standards for bicycle facilities reflecting research conducted at the University of California–Davis. The report was completed in 1975, ten years after Child wrote his letter to the editor. At that same time, a variety of booklets were written on bicycle facility design, including *Bikeways: Design and Construction Programs,* published by the National Recreation and Park Association (Jarrell 1974). Bicycle planning spread rapidly among U.S. cities during the 1960s and 1970s. Often, separate provisions were included for local errand-oriented bicyclists and touring bicyclists who favored speed on the roads.

Among bicyclists who rode five times a week, a national survey conducted in the 1970s indicated that 87 percent preferred sidewalk bikeways, 82 percent preferred bike lanes, 91 percent preferred separated bikeways that were not in parks, and 78 percent preferred signed routes, with people able to indicate a preference for more than one option. During the 1970s, the sidewalk bicycle paths and separated bikeways appeared to be favored over bicycling in the road or along signed routes. These preferences varied with individuals who bicycled for pleasure/exercise and those who bicycled to commute to work/school, but all bicycling groups had many travel corridor options that included the sidewalks, bike lanes, separated bikeways, signed routes, and regular roads (Kroll and Sommer 1976). In 1972, a young candidate, Dr. Dietmar Hahlweg, was elected mayor of Erlangen, Germany (Monheim 1990). A Fulbright scholar who had studied Jane Jacobs and Lewis Mumford at the University of Pittsburgh, he campaigned on the promise of providing urban-friendly transit in the historic hospital and university community. Rather than build highways, he fashioned innovative corridors including bikeways on sidewalks, through Woonerfs (streets closed to through traffic), on one-way streets, though parks, and on streets dedicated to buses, bicyclists, and pedestrians. Between 1974 and 1980, after stopping the road building and replacing it with public transportation and bicycle facilities, Erlangen had reduced the use of cars by 35 percent and had increased the use of bicycles by 26 percent 148).

In 1981, the American Association of State Highway and Transportation Officials wrote guidelines for the development of bicycle facilities. The primary components of the guidelines then and still are for bicycling in the road, with some text for distant leisure-based recreation paths. The guidelines do not include aesthetic components, such as adjacent greenery, or destination components, such as the desired location of human-need destinations, but instead focus on safety and engineering. Nonetheless, based on the preferences in communities including Davis, California, and Erlangen, Germany, community-wide systems of separated shared-use paths that connected to purposeful destinations such as stores were being built, tested, and successfully used.

In 1986, the Rails-to-Trails Conservancy was established to create a nationwide network of bike trails on former rail lines. In 1987, *The Report of the President's Commission on Americans Outdoors* suggested that communities should "establish Greenways, corridors of private and public recreation lands and waters, to provide people with access to open spaces close to where they live, and to link together rural and urban spaces in the American Landscape" (President's Commission on

Americans Outdoors 1987, 142). This report was further refined with the accepted principle that greenways should ideally be fifteen minutes from everyone's home. Based on these initiatives, greenways or linear parks were created across the United States, with many created on existing corridors such as railroad beds. Used primarily for recreation purposes during leisure time, these paths may or may not lead to useful destinations.

Deficiencies of Parks and Playgrounds and Design Considerations

The creation of parks in cities such as New York sought to provide open space (not necessarily "limited"), fresh air, sunlight, and alleviation of overcrowding. These parks included school playgrounds that provided opportunities for children to engage in physical activity before, during, and after school. The parks' greenery also provided psychological benefits and mentally restorative views, contributory elements in dense cities with few trees and gardens (Kaplan and Kaplan 1995). The adjacency of greenery in a park might lessen boundary pushing or delinquent behavior, as indicated in research on low-income African American boys. Young boys who lived near trees and parks also did better in school, had better peer relationships, and interacted better with their parents (Obasanjo 1998).

A park could also encourage community social interaction among diverse socioeconomic groups through the provision of urban design features or "social bridges" (Lusk 2002). "Social bridges" are characterized as assist, connect, observe, in absentia, or information. In an *assist* social bridge, someone helps someone else, as when a bicyclist steps aside while a novice in-line skater maneuvers a narrow bridge. A *connect* social bridge is based on Whyte's triangulation in which a third party or element can trigger conversation between two people, including strangers (Whyte 1980). The shared nostalgic environment of a porch, railings, and rocking chairs can foster a conversation between strangers who might not otherwise converse (figure 1). An *observe* social bridge occurs when a kindness is witnessed and humanity is affirmed, as in witnessing an adult helping a child learn to ride a bicycle. *In absentia* social bridges exist when the contribution of an absent party, perhaps the designer of the space, is implicitly acknowledged. For example, if a water element exists in a park that elicits laughter, gratitude is felt for the designer or the community members who provided it. In an *information* social bridge, information is imparted to the other person who might be absent but is present through language. In Paris, diseased historic trees had to be cut down in a park, but the park managers had written a sign that explained the reasons. Humanity was reaffirmed in this connection between an absent tree caretaker and the reader.

These parks, however, required travel time to reach them on foot and also the time to use the parks while there. Today, leisure time is a rare commodity, especially for low-income individuals who may have multiple jobs and limited time. With

Figure 1 Wraparound porch at restored train station on West Orange Trail near Orlando, Florida. (Photo by Anne Lusk.)

issues of crime, children may need to be supervised in public parks, and parents may not have the leisure time to be with their children, especially during daylight hours after school when single parents or dual-income parents are working.

The garden city parks as well as the sports fields were isolated from the shopping areas in Levittown. It was therefore difficult to combine playground time for children with shopping in the same trip. With sports facilities surrounded by residential homes, crime would be less of an issue in a close-knit community. Adult presence, however, is usually required to structure games, prevent bullying, and curtail playground or park damage.

Although recreation fields provide opportunities for physical activity, parents are often spectators rather than participants, specialized equipment is needed, a level of skill is required, and some school children feel excluded. Furthermore, team and field-based sports are often not continued in adulthood.

Rather than creating an isolated pocket park or sports field that primarily enhances the property values of adjacent properties, a design consideration might be to locate a grocery store next to a park or sports field to allow a parent to combine a utilitarian trip with a child's play. Children might be more motivated to travel to the grocery store, perhaps on foot, if they knew they were also going to be rewarded with time at the playground. If parks also are located close to where people live but include play structures and bicycle paths, the younger child could be monitored from home, as in Radburn, New Jersey, and the older child could travel a safe distance from home, with distance dependent on their age.

Deficiencies of Walking, Bicycling, and Skating Facilities

Although the existence of sidewalks for pedestrians and roads for bicyclists would suggest that the required thirty to ninety minutes of physical activity most days of the week could be achieved, a fine-grained analysis of these environments reveals design flaws. Pedestrians in community centers often do use sidewalks, but if parallel parking is available, many arrive by car and only walk a few feet to stores or eateries. Neighborhood sidewalks exist, but, especially in suburban tracts, they only circle homes and do not lead to a purposeful neighborhood grocery store. Children are often driven to school even though a sidewalk system might exist. Adults might use the sidewalks for a leisure time evening stroll, but leisure is a precious commodity and might not be spent on a walk.

Motor vehicles and driving habits have changed drastically since parents first let their children bicycle in the road. Although it is still possible to bicycle in the road, bicyclists who travel with vehicular traffic tend to be a certain age, a certain weight, and male with athletic abilities. The population that should engage in physical activity includes people of differing ages, with additional weight, males and females, and the less adroit. Sidewalks could be an option for bicycling or in-line skating, but pedestrian advocates now seek to ban all cycling and skating from sidewalks, leaving the slower bicyclists or skaters with no sanctioned place to travel except the distant leisure-based recreation path that does not lead to purposeful destinations.

Alexander, Lynch, and Whyte all mentioned the value of destinations as motivational components for physical activity. If the major destinations are in downtown community center stores, coffee shops, or movie theaters and there are policies banning bicycling or in-line skating on the sidewalks, the only people who can arrive at the destinations are car drivers, transit riders, pedestrians, and bicyclists capable of bicycling on the road. Even though slower bicyclists and in-line skaters could walk the distance, either with their bike or after locking their bike or by carrying shoes in a backpack, sometimes the distances are too great to walk in spread-out downtowns.

Physical activity cannot currently be a routine part of the day for "all" populations because only pedestrians and road bicyclists can arrive at the purposeful destinations such as a grocery store. The first design option would be to provide new urban forms, or European cycle tracks, in downtown community centers so that slower bicyclists and skaters could arrive at key destinations. Created either as part of the sidewalk or a curb step down from the sidewalk but between parallel parked cars or traffic, the cycle tracks could separate pedestrians from other users

Figure 2   European cycle track for bicyclists in Paris, France. (Photo by Anne Lusk.)

considered wheeled pedestrians (figure 2). Combined with separate shared-use paths and residential streets, a grid could be developed that would enable the physically inactive to bicycle, jog, or skate to preferred destinations, especially to downtown community destinations such as stores, coffee shops, and the post office. The cycle tracks could also be used as part of the Safe Routes to School program on specific streets for safe travel to and from school.

A second urban form design option would be to put purposeful destinations, such as grocery stores, on recreation paths. Existing corridors such as shared-use paths or greenways could have useful destinations added to the corridor. Rather than bicycle only on Saturday when leisure time might be available, routine trips could be made to the grocery store or drugstore located adjacent to a greenway. Research on six of the most preferred greenways in the United States showed that habitual users identified a mean of 3.17 destinations (SD 1.32); the mean distance between destinations was 3.92 miles (SD 2.65, variation due to user type); and the means at the destinations were 46 features, 8.1 activities, and 14.6 meanings. On the Chicago Lakefront Trail where a total of forty-one destinations were identified, individuals still identified three to four destinations. The observations at the destinations indicated that people stop at some destinations, characterized as "social stop." On the Stowe (Vermont) Recreation Path, users commonly stopped at the farmers' market set up each Sunday in a field adjacent to the path where they socialized with friends and also purchased fresh produce (figure 3). Other destina-

Figure 3   Farmers' market destination on the Stowe Recreation Path. (Photo by Anne Lusk.)

tions, characterized as "positive-identity pass-by," exist where people pass by. At theses destinations, the participants would add vocabulary words to the survey forms with a positive reflection. For example, someone wrote, "Doggie Beach where I enjoy seeing the dogs play." Observations at this destination showed that few people actually stopped, although they did look in the direction of the dog beach. Further analysis of the data suggested that some destinations, such as Confluence Park in Denver, were "prowess plazas" that showcased healthy athleticism as kayakers maneuvered the rapids in the South Platte River.

The above two sections are the most significant piece in making the case for new urban forms. We need the European cycle tracks and grocery stores on the leisure-based recreation paths instead of just sidewalks and roads for bicycling. We have lost much ground since the 1970s when people could bicycle on the sidewalks. The literature in the 1980s focused on leisure-based parks and play, not physical activity as a routine part of the day. Not all populations have the leisure time, especially populations suffering the most from obesity.

Design Considerations for the Humane Metropolis

It was not with malice that parks, such as Central Park, were created distant from the low-income population who had only Sunday as a day of rest. The future would bring a variety of affordable transportation forms for all populations to arrive at

Central Park. Parks were intentionally separated from commerce because of the teams of wagons and horses delivering goods to the stores and pollution in association with trades such as the slaughtering yards in Chicago. People were not overweight because of diet and the daily involvement of physical work or mandatory exercise such as walking for transportation, so routine physical exercise for the sake of health was not necessary. For bicycle facilities, the most audible champions have been road bicyclists who choose speed on the road over the chaos of a shared-use path. Six-year-old children and sixty-five-year-old nonbicyclist women have not been invited participants in transportation meetings about urban forms. All the while, rates of obesity increased as did associated diseases.

The humane metropolis requires parks with nearby purposeful amenities such as grocery stores. Rather than isolated pocket parks with only the benefits of greenery, the parks could also be linear and connect with other parks or corridors, facilitating use near home and travel to other locations. Parks, especially with play features, could be located near stores to combine utility with leisure. Safe roads for bicycling, sidewalks for walking, and European cycle tracks could provide all populations with access to important community destinations. Residential areas with less traffic could allow bicycling or in-line skating on the sidewalks or feature traffic-calming devices, such as Woonerfs or semiclosed roads, to permit safe passage on the road for less skilled bicyclists and skaters. Separate, dedicated shared-use greenways could offer human-need destinations, including the places frequented as part of trip chaining in a car, such as grocery stores, banks, and coffee shops.

Destinations could also feature "prowess plazas" and showcase athleticism with an expanded list of culturally inclusive activities, such as basketball, jump rope, and skateboarding Thus, young athletic stars would be witnessed and discussed by neighbors, and "social bridges" among spectators and participants would be enhanced. The design would focus less on the car and more on the people, their health, and interaction with one another.

"Health enterprise zones" could help financially foster the establishment of produce stores, gyms, and other health and fitness-oriented businesses in rundown neighborhood districts or in the vicinity of linear greenways. Created as a healthful form of an economic enterprise zone but with similar objectives of fostering community strengths and building social capital, all businesses encouraged in the health enterprise zone would have to in some way benefit health. Fast-food chains, quick stops, liquor stores, bars, billboards selling unhealthy food, or vending machines with sugar sodas could be banned within the health enterprise zone, whereas special incentives could be offered to attract health-inducing businesses. The creation of the health enterprise zone would change the buying and membership opportunities of neighborhood residents and also raise awareness of the benefits in different types of establishments, foods, and activities. Connected to or within this zone could be new parks, European cycle tracks, or greenways that all enabled physical activity as a routine part of the day.

In some states, mandatory provision of land for a bikeway or play space may be required of subdivision developers (Platt 2004, 276–78). Building on that precedent, the development approval process may be amended to require "health impact analysis" of proposed development not currently considered in building permit reviews. Such provisions might require the proposed plan to include features that encourage outdoor exercise other than simply walking between the premises and a parked car. A European cycle track could allow for passage to the restaurant by bicycling or in-line skating. Food and beverage consumption could then be balanced by energy burning.

The more humane metropolis should be responsive to the health crisis and should promote routine physical activity for all populations. The needs of all segments of the population, including the elderly, the physically challenged, youth, and nondrivers, should be served through responsive urban forms. Above all, the promotion of fitness and health should be integral to new development and should be added through retrofitting existing parks, streets, recreation paths, shopping areas, downtowns, and residential districts. In addition to diet, physical inactivity is responsible for the current high levels of obesity and resultant health problems. Health-oriented community design could be the twenty-first-century public health equivalent to progressive reformer Edwin Chadwick's studies in the 1830s of sanitary conditions in the industrial slums of England and could help reverse this growing national epidemic of obesity.

References

Alexander, C., S. Ishikawa, and M. Silverstein. 1977. *A pattern language: Towns-buildings-construction.* New York: Oxford University Press.

Alland, A., Sr. 1975. *Jacob A. Riis photographer and citizen.* London: Gordon Fraser.

Appleyard, D., K. Lynch, and J. R. Myer. 1966. *The view from the road.* Cambridge, MA: MIT Press.

Caro, R. A. 1984. *The power broker: Robert Moses and the fall of New York.* New York: Knopf.

Centers for Disease Control and Prevention. 2002. Behavioral risk factor surveillance system, surveillance for certain health behaviors among selected local areas—United States, behavioral risk factor surveillance system, 2002. *Morbidity and Mortality Weekly Report* 53 (No. SS-05).

———. 2004. Youth risk behavior surveillance system, youth risk behavior surveillance—United States, 2003. *Morbidity and Mortality Weekly Report* 53(No. SS-2): 1–29.

Department of Health and Human Services and Department of Agriculture. 2005. *Dietary guidelines for Americans 2005.* Available online at 222.health.gov/dietaryguidelines/dga2005/document/html/chapter4.htm.

Fisher, I. D. 1986. *Frederick Law Olmsted and the city planning movement in the United States.* Ann Arbor: University of Michigan Research Press.

Handy, S. L., M. G. Boanet, R. Ewing, and R. E. Killingsworth. 2002. How the built environment affects physical activity: Views from urban planning. *American Journal of Preventive Medicine* 23(2S): 64–73.

Jarrell, T. R. 1974. *Bikeways: Design-construction-programs.* Arlington, VA: National Recreation and Park Association.

Kahn, E. B., L. T. Ramsey, R. C. Brownson, G. W. Heath, E. H. Howse, K. E. Powell, E. J. Stone, M. W. Rajab, and P. Corso. 2002. The effectiveness of interventions to increase physical activity. A systematic review. *American Journal of Preventive Medicine* 22(4) (Supp. 1): 73–107.

Kaplan, R., and S. Kaplan. 1995. *The experience of nature: A psychological perspective.* Ann Arbor, MI: Ulrich's Bookstore.

Killingsworth, R., J. Earp, and R. Moore. 2003. Supporting health through design: Challenges and opportunities. *American Journal of Health Promotion* 18:1–3.

Kroll, B., and R. Sommer. 1976. Bicyclists' response to urban bikeways. *AIP Journal*, January, 42–51.

Lott, D. 2003. How our bike lanes were born: A determined group of Davis activists just wouldn't give up on our quality of life. Special to the *Davis Enterprise.* Available online at www.runmuki.com/paul/writing/lottarticle.html.

Lusk, A. C. 2002. Guidelines for greenways: Determining the distance to, features of, and human needs met by destinations on multi-use corridors. Ph.D. diss., Taubman College of Architecture and Urban Planning, University of Michigan–Ann Arbor.

Lynch, K. 1960. *The image of the city.* Cambridge, MA: MIT Press.

Monheim, R. 1990. Policy issues in promoting the green modes. In *The greening of urban transport: Planning for walking and cycling in Western cities,* ed. R. Tolley, 148. London: Belhaven Press.

Moudon, A. 1987. *Public streets for public use.* New York: Van Nostrand Reinhold.

Mumford, L. 1965. The guarded city idea and modern planning. In E. Howard, *Garden cities of tomorrow,* 29–40. Cambridge, MA: MIT Press.

Obasanjo, O. O. 1998. The impact of the physical environment on adolescents in the inner city. Ph.D. diss., University of Michigan–Ann Arbor.

Olmsted, F. L. 1865. *The value and care of parks.* Reading, MA: Addison-Wesley.

———. 1868. *Preliminary report upon the proposed suburban village, near Chicago.* New York: Sutton, Brown.

Page, M. 1999. *The creative destruction of Manhattan, 1900–1940.* Chicago: University of Chicago Press.

Pate, R. R., et al. 1995. Physical activity and public health: A recommendation from the Centers for Disease Control and Prevention and the American College of Sports Medicine. *Journal of the American Medical Association* 273(5): 402–7.

Perry, C. 1929. *A plan for New York and its environs.* New York: New York Regional Planning Association.

Platt, R. H. 2004. *Land use and society: Geography, law, and public policy.* Rev. ed. Washington, DC: Island Press.

President's Commission on Americans Outdoors. 1987. *Americans outdoors: The legacy, the challenge.* Washington, DC: Island Press.

Rybczynski, W. 1999. *A clearing in the distance.* New York: Scribner's.

Sallis, J. F., K. Kraft, and L. S. Linton. 2002. How the environment shapes physical activity: A transdisciplinary research agenda. *American Journal of Preventive Medicine* 22(3): 208.

Scott, M. 1969. *American city planning since 1890.* Berkeley: University of California Press.

U.S. Department of Health and Human Services. 1996. *The effects of physical activity on health and disease: A report of the Surgeon General.* Washington, DC: U.S. Department of Health and Human Services.

————. 2000. *Healthy people 2010,* 2nd ed. *With understanding and improving health and objectives for improving health* (2 vols.). Washington, DC: U.S. Department of Health and Human Services.

Ward, S. V., ed. 1992. *The garden city introduced.* In The garden city: Past, present, and future, 1–27. London: E & FN.

Whyte, W. H. 1968. *The last landscape.* New York: Doubleday.

————. 1980. *The social life of small urban spaces.* Washington, DC: Conservation Foundation.

A Metropolitan New York Biosphere Reserve?

William D. Solecki and Cynthia Rosenzweig

By 2025, it is estimated that five billion of the earth's total population of eight billion people will live in urban settlements (United Nations 1995). Urban environments involve complex and intense interaction between ecological and human systems at various geographic scales from the neighborhood to the megalopolis. Yet even though natural functions and phenomena are greatly transformed by urban development, they are not eradicated. Indeed, urban places retain many vestiges of ecological functions and services. For example, coastal wetlands in urbanized settings simultaneously provide areas for active and passive recreation, spawning ground for regional fisheries, places for water quality control, and stopover points for migrating birds.

It is now recognized that the relationship of human and natural systems is in constant flux (see Haughton and Hunter 1994; Platt, Rowntree, and Muick 1994; Bennett and Teague 1999). Urban ecological systems are projected to become even more dynamic in the future, particularly as a result of global climate change. This issue is creating a new relationship between the global scale and local places (Kates and Wilbanks 2003). By the end of the twenty-first century, for example, global climate-related increases in sea-level rise in the New York City region could be up to four times greater than the current rate of rise occurring naturally, which would dramatically affect coastal ecosystems (Gornitz 2001).

The biosphere reserve (BR) concept, as formulated by the UNESCO Man and Biosphere (MAB) program (http://www.unesco.org/mab/), is an approach to regional environmental management that attempts to foster a set of goals, including biodiversity protection, long-term environment monitoring, and sustainability modeling (Batisse 1993). The origin of biosphere reserves was the Biosphere Conference organized by UNESCO in 1968, the first intergovernmental conference to seek to reconcile the conservation and use of natural resources, thereby foreshadowing the present-day notion of sustainable development. The aim of the conference was to establish terrestrial and coastal areas representing the main ecosystems of the planet in which genetic resources would be protected and where research on ecosystems as well as monitoring and training work could be carried out for an intergovernmental program. UNESCO officially launched the MAB program in 1970. One of the program's projects was to establish a coordinated world network of new protected areas, to be designated as biosphere reserves, in reference to the program itself.

The World Network of Biosphere Reserves, now numbering more than four hundred sites, is formally constituted by a statutory framework that resulted from the work of the International Conference on Biosphere Reserves held in Seville, Spain, in March 1995. This statutory framework sets out ground rules of the network and foresees a periodic review of biosphere reserves. Activities of the network are guided by the Seville Strategy for Biosphere Reserves, also drawn up at the Seville conference. At present, not all existing biosphere reserves fully participate in the network, and the goal of these guiding documents is to help improve their functioning in the coming years.

Long conceived as a program for managing locations with high levels of biodiversity in nonurban settings, the MAB program has evolved to include the notion of management of biodiversity within urban places. The Columbia University/ UNESCO Joint Program on Biosphere and Society (CUBES) is one effort that has fostered this extension of the biosphere reserve concept (http://earthinstitute .columbia.edu/cubes/). In Cape Town, South Africa, and New York City, CUBES has built a network of urban biosphere groups composed of researchers, policy specialists, and municipal and governmental officials, and in 2003 it hosted the Urban Biosphere and Society Conference as part of its mission to develop sustainable networks of cooperation to support globally relevant local strategies for poverty alleviation, environmental sustainability, social inclusion, and conflict mitigation.

The CUBES activities in New York, fostered discussion of how an urban biosphere reserve might be created in a global city in general and in New York City specifically. The aim of this essay is thus to assess the potential value of the biosphere reserve concept as applied within the greater New York metropolitan region. Although there are many possible ways to delimit this area geographically, this essay will use the Regional Plan Association definition of a thirty-one-county area, lying within the states of New York, Connecticut, and New Jersey, with a 2000 population of approximately 21.5 million.

Urban and Regional Environmental Management

Throughout the historical development of cities, the role and function of nature typically has been defined within the context of its social utility or function. In the United States in the mid-nineteenth century, the importance of promoting nature in cities first emerged with the recognition that parks can provide the social function of enabling urban populations to find relief from the congestion of city life. This goal motivated the work of Frederick Law Olmsted and the development of Central Park and other greenspaces as "green lungs of the city" (Cranz 1982; Burrows and Wallace 1999). More than a century later, the environmental functions of natural areas in urban areas have become recognized as well. Natural areas provide ecological services, such as the following:

- Air quality enhancement (e.g., trees and other vegetation promote cleaner air via pollutant removal)
- Flood protection (e.g., wetlands act as stormwater catchment areas)
- Urban heat island) abatement (e.g., trees help moderate daytime and night-time temperature increases via shade and evapotranspiration)
- Water quality protection (e.g., stream-corridor vegetation prevent siltation).

With increasing recognition of natural area function in urban areas, a significant debate, focusing largely on the size of the parcels and the species composition, has emerged regarding the quality and characteristics of the nature at these sites (Beatley 1994). Several fundamental questions have been raised. For example, how should one establish the environmental value of vastly different-sized parcels, ranging from large city parks such as Central Park to small patches of ground in front of buildings? What species or ecological functions should be the focus of ecological and environmental planners? Should the focus be on all species (including invasive or "alien" species) and functions (e.g., active recreation) found within cities, or just on native species and those that provide "natural" (e.g., passive recreation) functions?

Underlining these questions has been improved understanding of urban ecological and environmental function. Natural systems of highly urbanized places are clearly altered. Ecologists and planners, however, are now recognizing that remnant natural areas may still provide ecological and environmental services. Studies have shown some cities to be richer, biologically speaking, than surrounding suburbs and agricultural areas (Savard, Clergeau, and Mennechez 2000). Increased focus on system-level understanding of the ecology, biodiversity, and environment of urban areas has helped foster the rise of a new wave of regional environmentalism in metropolitan areas (Taylor and Hollander 2003). The following are some examples of programs or strategies in progress in the United States today:

- *Long-term ecological research (LTER) sites for urban areas (http://lternet.edu/).* The LTER program, sponsored by the National Science Foundation, focused on ecological inventory, long-term monitoring, assessment, and research-carried out by public agencies and academic institutions in a cooperative administrative structure at twenty-four specific sites. Two urban LTER sites—Baltimore, Maryland, and Phoenix, Arizona—were designated in the late 1990s. The urban LTER sites have a key function of assessing the impacts of human-environment interactions on ecological processes.
- *U.S. National Estuary Program (http://www.epa.gov/nep/).* This program was established by the U.S. Congress in 1987 to improve the quality of estuaries of national importance. The U.S. Environmental Protection Agency directs the development of plans for attaining or maintaining water quality of estuar-

ies for use as a drinking water source, indigenous species habitat, and recreation resource. Twenty-eight national estuaries have been defined, several of which—Puget Sound, San Francisco Bay, Galveston Bay, and New York/New Jersey Harbor—are in highly urbanized sites.

- *Interagency federal task forces.* Numerous federal interagency task forces were created, particularly during the Clinton administration, to address specific regional environmental issues (e.g., ecosystem restoration of the Florida Everglades). The principal goals of these programs have been to initiate conditions through which the current status and long-term patterns of ecological dynamics of region could be studied by bringing together the suite of federal and other government scientists and stakeholders together. The research brings together analysts from the biological, physical, and social sciences to collect data and synthesize existing information on how the ecological and engineered systems of the region work. These results could be used to generate policy proposals.

- *Regional networks and alliances.* These programs represent an emerging organizational structure that attempts to loosely link many environment-focused institutions in a single area or region and direct them toward a set of common goals and objectives. The Chicago Wilderness is the most prominent urban nature alliance. Chicago Wilderness is defined as "a partnership of more than 180 public and private organizations that have joined together to protect, restore and manage natural lands around Chicago and the plants and animals that inhabit them. The goals of the partnership are to help restore natural communities on public and private lands; prevent the ongoing loss of critical habitat and promoting careful development, and provide opportunities for citizens to become involved in local biodiversity conservation" (www .chicagowilderness.org). This overarching structure is creating a new definition of regional environmental consciousness and citizenship for the Chicago region. Such a program, however, lacks consideration of linkages to larger scales and more contested issues, such as global climate change and environmental justice.

- *Specific watershed-based nongovernmental organizations.* Some local nongovernmental organizations focus on the ecological character and functioning of specific watershed and associated rivers. Many such watershed organizations are present in highly urbanized areas. Although scientific research is occasionally part of their mandate, they are often more focused on the achievement of a specific conservation goal, such as protection of open space or the enhancement of native wildlife in a given area (Cortner and Moote 1999). Some of these initiatives are rather diffuse and are largely presented as planning proposals with more general goals, such as protecting biodiversity, rather than as the foundations of comprehensive new public policies.

The Biosphere Reserve Concept: Promise and Limitations

Another model of regional environmental management is the biosphere reserve concept. This concept has been put into practice at more than four hundred sites in more than ninety countries since 1971 (UNESCO 1996). As of 2003, there were forty-seven designated sites in the United States. To date, the biosphere reserve concept has mostly been applied to wilderness or rural sites away from major settlements, although several reserves already exist in urban fringe locations. For example, the Golden Gate Biosphere Reserve includes thirteen protected areas in the greater San Francisco Bay area, the Everglades National Park borders metropolitan Miami, and the Pinelands National Reserve in southern New Jersey is surrounded by urban and suburban development on three sides. The application of the concept to urban areas is formally under review by the MAB program (UNESCO 2000), and there is continuing lively discussion on the MAB Urban Group Forum (http://www.unesco.org/cgi-ubb/forumdisplay.cgi). Koichiro Matsuura, director-general of UNESCO on the occasion of World Environment Day, June 5, 2005, reported that Canberra, Cape Town, Istanbul, and Rome are actively exploring the application of the UNESCO biosphere reserve concept and that the São Paulo City Green Belt Biosphere Reserve promotes eco-job training for young, poor urban people, covering topics such as water and waste management, recycling, and ecotourism.

As implemented by UNESCO-MAB (UNESCO 2003), biosphere reserves are intended to serve three primary functions that are complementary and mutually reinforcing

- A *conservation function* to contribute to the conservation of landscapes, ecosystems, species, and genetic variation
- A *development function* to foster economic and human development that is socioculturally and ecologically sustainable
- A *logistic function* to provide support for research, monitoring, education, and information exchange related to local, national and global issues of conservation and development.

These goals can be achieved through a diversity of strategies (Alfsen-Norodom and Lane 2002; Bridgewater 2002). A common strategy has been the demarcation of core, buffer, and transition management zones. The *core* of the reserve encompasses the critical habitats and biodiversity resources to be protected. The role of the *buffer* area is to protect the core, and the *transition* area serves as an intermediate zone between the buffer and the surrounding territory. Relatively few of the existing biosphere reserves are composed literally of three concentric rings of management areas. In most cases, the core is not a single site nor is it completely surrounded by a buffer zone. The core areas are often a set of parcels with the most exclusive zoning restrictions. In the New Jersey Pinelands, for example, the core pinelands protection areas are scattered throughout the central part of the reserve.

Although biosphere reserves encompass very diverse types of landscape, the following criteria usually must be met to quality for designation by UNESCO-MAB (UNESCO 2003):

- Be representative of a major biogeographic region, including a gradation of human intervention
- Contain landscapes, ecosystems, or animal and plant species, or varieties that need to be conserved
- Provide an opportunity to explore and demonstrate approaches to sustainable development within the larger region where they are located
- Be of an appropriate size to serve the three biosphere reserve functions mentioned above
- Have an appropriate zoning system with a legally constituted core area or areas, devoted to long-term protection; a clearly identified buffer zone or zones; and an outer transition area
- Have a management structure to involve all stakeholders, including relevant public authorities, local communities, and private interests.

National MAB committees are responsible for preparing biosphere reserve nominations and for involving the appropriate government agencies, institutions, and local interests in preparing the nomination. Each nomination is examined by the UNESCO Biosphere Reserve Advisory Committee, which in turn makes a recommendation to the International Coordinating Council of the MAB program. The council makes the final decision on designation (UNESCO 2003).

Although designation of a biosphere reserve may reinforce environmental protection strategies already existing or proposed for the site in question, it does not involve any mandatory management constraints. There is no of loss of sovereignty over the site to the United Nations or to any other international body (UNESCO 1996).

Still, the implementation of the biosphere reserve concept has provoked controversy. Concerns in periurban areas have focused on the effect of BR designation on economic shifts and declines in traditional industries (e.g., extractive industries such as mining), changes in development patterns (e.g., increased growth of largely low-wage, service-based industries such as tourism), and limitations in institutional capacity of local communities (e.g., hamlets and villages inability to respond to the service needs of second-home, seasonal residents) (Solecki 1994). Another critique is that BR planning has often been a "top-down" process in which local stakeholders have little influence. In recent years, there has been an explicit effort by UNESCO-MAB and the various national MAB committees (e.g., U.S.-MAB, Canada-MAB) to incorporate local stakeholders in the biosphere reserve process.

The mission and functions of biosphere reserves were reexamined at a 1995 UNESCO-MAB conference held in Seville, Spain (UNESCO 1996). The resulting Seville strategy defined a set of mechanisms through which a revised BR strategy could achieve not only the long-held conservation goals of reserves but also carry on research, long-term monitoring, training, and education, while enabling local communities to become more fully involved in the process. There has since been extensive discussion regarding the extension of the BR concept to urban settings where environmental and social issues are often intertwined and particularly acute (e.g., lack of green space in lower socioeconomic neighborhoods) (UNESCO Advisory Committee for Biosphere Reserves 1998). Douglas and Fox (2000) suggest that the concept could integrate multiple environmental initiatives and designations typically found in urban metropolitan regions. For example, BR designation might facilitate the development of new interagency agreement or other similar administrative mechanism, which could foster better connections among existing open space management areas currently managed by different levels of governments (e.g., city parks, county recreation areas, and state and federal wildlife sanctuaries).

Based on a review of experience with biosphere reserves in the United Kingdom, Price, MacDonald, and Nuttall (1999) urge that nominations of potential sites should arise from local communities and other stakeholders. According to Frost (2001, 7): "The overall goal of any new reserve must be to conserve nature by re-connecting people to it and helping them learn more about it, and so contribute to managing it in a sustainable way. It is possible to foresee a day when local communities will campaign for their areas to be designated as biosphere reserves in the same way that communities campaign 'World Heritage' status." Frost (2001, 218) proposes that *urban* biosphere reserves should

1. Be created at the request of and with support from local communities and key stakeholders
2. Have more than one area that is at least of special area of conservation, special protection area, or national nature reserve standard
3. Use local nature reserves, country parks, and local natural areas as buffer zones
4. Draw in the other elements of the urban area's network of open space as transition zones, which might include informal open space, industrial landscaping schemes, transport corridors, elements of the urban forest, and private open space
5. Have a management plan and planning mechanism that integrates the various local and environmental plans across administrative boundaries
6. Maintain stakeholder participation through the use of participatory techniques

7. Involve local education and research establishments in work to monitor and develop all aspects of the reserve, both human and environmental

8. Use the presence of the reserve to create a general focus on sustainability that informs decisions at all levels

9. Continue outreach work to bring all sections of the local community into contact with the reserve to enjoy, and be aware of nature within their daily life.

Urban biosphere reserves, like conventional ones, should provide biodiversity protection, long-term ecological monitoring, and sustainability experimentation and planning. The urban context, however, provides opportunity to examine the interactions of those three functions and their associated effects further. Any measure or indicator of success with respect to any one of these functions must be defined with respect to the other two. This intersection further reflects the spirit and challenge of the Seville strategy and gives urban biosphere reserves a special niche in the spectrum of regional environmental initiatives and planning concepts used in major metropolitan cities. Urban BRs have the potential to incorporate the *biodiversity protection* mission of efforts like Chicago Wilderness and the long-term *ecological monitoring* associated with the Baltimore and Phoenix LTER programs, as well as the BR sustainability function, such as resource use reduction initiatives, currently not developed in these types of initiatives. Some elements of urban BRs such as greenways (sites for wildlife habitat, instrumentation deployment, and urban heat island mitigation) can promote activities associated with all three missions. .

Urban Biosphere Reserve Sites

The integration of the biosphere reserve concept into the urban landscape is underway in several cities throughout the world. Current sites are found in São Paulo, Brazil; Arganeraie, Morocco; California's San Francisco Bay area; Kristianstad, Sweden; Rome, Italy; Cape Town, South Africa; and Melbourne and the Greater Canberra region in Australia. The Chicago Wilderness network is also considering applying for BR status (Sholtes 2003) for the Chicago area, and a movement has developed in Turkey to have an urban biosphere designated there (Matsuura 2005). These efforts take a variety of forms. Some include discrete sites, such as critical habitat or recreation areas, within the city or metropolitan region, whereas others include larger sections of urbanized zone. In several cases, existing urban biosphere reserves have become the impetus for more integrated, regional BR planning efforts (Matsuura 2005), as in the current planning efforts in Rome and Cape Town. In the San Francisco Bay area the Golden Gate Biosphere Reserve was created in 1988. This urban core reserve is connected via an administrative partnership to twelve other protected areas throughout the greater San Francisco Bay area (Golden Gate Biosphere Reserve Association 2005). The biosphere reserve effort in São Paulo metropolitan region represents one of the most spatially comprehensive set of activities in an urbanized area. The Sao Paulo City Green Belt Biosphere Reserve was recognized in 1994 as part of the larger Atlantic Forest Biosphere Reserve in Brazil. It comprises seventy-three municipal districts within the São Paulo state, including the capital city of São Paulo. The reserve includes critical conservation areas that are seen as providing ecological services such as water supply protection, regional thermal stabilization, hillside and flood control, and recreational amenities for the region's nineteen million inhabitants (Pires et al. 2002; São Paulo Forest Institute 2004). The reserve came into being after protracted negotiations between local nongovernmental organizations, city and state governments, and eventually federal government as well. Since the mid-1990s, its existence has enabled the development of sprawl control and other regionwide management proposals (São Paulo Forest Institute 2004). Local and state officials also have begun to investigate other economic benefits the reserve represents, such as a locus for sustainable ecotourism (Pires et al. 2002).

In Rome, interest in the biosphere reserve concept originated in the 1970s with studies to integrate ecological management strategies appropriate for urban processes and phenomena and to understand the interdependence between ecological systems (e.g., hydrologic, biotic) and the human systems (transportation, energy supply) present within Rome (Bonnes 1991, 1993). In response to findings of this research, Rome city officials endorsed proposing a biosphere reserve comprising the major greenspace areas of Rome. The resulting proposal was designed to (1) outline the development of the MAB-Rome project with a focus on "natural environment" management issues; (2) promote mobilization through collaboration and partnership among environmental decision makers, researchers, and citizens; and (3) apply the Seville strategy (Bonnes 2000; Bonnes et al. 2003).

Melbourne in 2002 decided to nominate a portion of Mornington Peninsula as an urban biosphere reserve (Miller 2002). The 2,100-square-mile reserve is designed to be a model for how sustainable development could be implemented within a relatively urban setting. The reserve contains a population of about 200,000 and is situated on the suburban fringe of Melbourne. Local government agencies and the community regard the biosphere as a catalyst for bringing public and private sectors together to generate a shared vision for the city. Businesses with green practices plan to use the reserve as a marketing tool. Local officials believe that it will become a center for the study of ecologically sustainable development and urban planning (Miller 2002).

Also in Australia, a coalition of organizations has nominated the Australian Capital Territory and Queanbeyan City in Canberra as a single biosphere reserve. Defined as the "bush capital," the Canberra reserve would help promote existing environmental protection activities in the region, provide a mechanism to voice community interests, and showcase the local biocultural regions to generate regional and national pride, encourage tourism, and promote local products (Australian National Sustainability Centre 2003).

In the Cape Town metropolitan region, the Capetown Urban Biosphere Group is examining how the BR concept may be used in an urban African context for the development of environmental conservation and management programs yielding sustained benefits to the poor (Stanvliet et al. 2004). The geographical focus is on the Cape Flats area as the potential overlapping transition zone for four biosphere reserves making up the proposed Cape Town biosphere reserve cluster. Two of the areas are currently biosphere reserves: Kogelberg and West Coast. The other two areas are proposed reserves: Boland and Table Mountain–Peninsula Chain. Several hundred thousand people live in the reserves as permanent or semipermanent residents. The core of each reserve is composed of important ecological sites with endemic species. In Cape Town, the reserves are seen as valuable for environmental education and ecotourism as well as for critical habitat.

The New York Metropolitan Region: Framework of Environmental Management

The New York Metropolitan Region (NYMR) as defined by the Regional Plan Association encompasses thirty-one counties in three states with a total population of about 21.5 million, of which nearly one-third live in New York City. New York City has a population density of about 10,204 people per square kilometer, as compared with only 422 people per square kilometer for the rest of the region. Jurisdictionally fragmented, in addition to the thirty-one counties, there are some 1,600 cities, towns, and villages in New York, New Jersey, and Connecticut, besides the federal government and several regional organizations (Zimmerman and Cusker 2001).

With 2,413 kilometers (1,500 miles) of tidal shoreline, the region's development has been intimately connected to the ocean (Burrows and Wallace 1999). Four of the five New York City boroughs are located on islands (Brooklyn, Manhattan, Queens, and Staten Island). Large waterways and bays, including the Hudson River, East River, Long Island Sound, Peconic Bay, Jamaica Bay, and the Hackensack Meadowlands, cut deeply into the land area. Given its coastal location, much of the land area is at relatively low elevation. About 1 percent of the land is below three meters in elevation. This 1 percent encompasses some of the most heavily developed land and regionally important infrastructure in the NYMR.

Beginning with the consolidation of its five boroughs in 1898, the NYMR emerged as one of the world's primary economic, cultural, and educational centers. Economic and social changes were reflected in ever greater disparities of wealth and poverty, high expanding sociocultural diversity, a diminishing stock of affordable housing, and rapid suburbanization but continued dominance and gentrification of the urban core. Despite its power and wealth, the region's social, economic, and environmental fabric is at risk (Yaro and Hiss 1996). The dominant ecological and economic trends have made the region more vulnerable to perturbations such as extreme coastal storms, less sustainable as a result of rising demands on regional resources, and less equitable in terms of pollution exposure and quality of life.

Many of these processes directly result from societal interaction with the natural systems of the region. For example, the region's coastal location makes the highly developed, nearshore areas vulnerable to coastal storms and sea-level rise. The physiography of the region's main river basins (e.g., lower Hudson, Passaic, Raritan, and Hackensack rivers) tends to concentrate pollutants in the densely settled estuarine area of the region (Tarr and Ayres 1990). In addition, suburban sprawl is simultaneously straining local water supplies and increasingly threatening the quality of regional water supply systems, as evidenced by growth in development and associated pollutant runoff in the Croton and trans-Hudson source watersheds serving New York City (National Research Council 2000) as well as suburban development around the Jersey City and Newark water supply catchment areas in New Jersey.

The majority of the region's natural historic wetlands have been lost, and buffer areas around wetlands or rivers typically no longer exist (Hartig et al., 2002). In many areas, smaller rivers and streams have been filled, channelized, or placed into culverts. Surface water and groundwater supplies, particularly in the more heavily urbanized areas, have been compromised and typically exceed federal water pollution standards. There are more than 100,000 leaking underground fuel tanks, spill sites, and former industrial sites included on the federal government's register of known or potential toxic sites (Yaro and Hiss 1996). Many are located in lowland locations where coastal wetlands were historically used as landfill sites.

Even with this history of degradation, the few remaining habitat sites in the region provide critical ecological functions, such as stopping points for migratory bird species. As species have adapted to human disturbances in recent decades, species richness has improved in such major habitat preserves as New Jersey's Hackensack Meadowlands and Great Swamp, and New York's Jamaica Bay has become cleaner (Waldman 1999; New York/New Jersey Harbor Estuary Program 2002).

Environmental management is politically fragmented in the NYMR (Solecki and Shelley 1996). Since the mid-1990s, however, there has been increasing interest in using regional frameworks to address multijurisdictional environmental problems, such as water quality and supply, and open space and biodiversity protection. Some of these regional identifiers have emerged out of federal programs; others have been developed by nongovernmental organizations. These efforts emerged from three different organizing elements:

1. *Administrative.* Conservation activities are centered on existing special management areas such as New Jersey's Hackensack Meadowlands, New Jersey Highlands, New York State Long Island Pine Barrens, and New York City's Jamaica Bay. These areas often represent remnants of larger ecological zones degraded by land use and land cover change that are now protected via zoning or public ownership.

2. *Hydrological.* Conservation activities are focused on watersheds as the primary unit of analysis and management. Examples include the New Jersey's watershed management strategy; the New York/New Jersey Harbor Estuary Program (http://www.feagrant.sunysb.edu/hep/), which is part of the U.S. EPA National Estuary Program; and the NGO-driven Highlands to Ocean (H_2O) initiative designed to promote open space and habitat protection within the region draining to New York's harbor and estuary (http://www.rpa.org/projects/openspace/) and which is promoted by the Regional Planning Association/Highlands Coalition.

3. *Metropolitan region.* Conservation activities are centered on the larger New York Metropolitan Region, which typically includes a region some 100 to 150 kilometers from New York City. Current activities include the Wildlife Conservation Society Bioscape Program (http://www.nybioscape.org/) and the Nature Network, an incipient alliance of environmental institutions and agencies in the New York area.

A Metropolitan New York Biosphere Reserve?

To help remedy ecosystem degradation, unite ongoing local efforts, and promote more effective long-term ecological and economic patterns of consumption and greater resilience to future perturbations, one or more biosphere reserves within the New York metropolitan area should be considered. Applying the concept in the region could benefit the local ecology and economy. The socioeconomic benefits of BR planning, such as enhanced water supply protection, flood control, and better recreational amenities, will flow from the improved environmental function of the region's open space and other sites where natural systems function is evident.

Any proposed urban biosphere reserve must reflect the Seville strategy guidelines for reserve development and structure. One crucial element is the inclusion of a wide spectrum of relevant local groups and organizations into the development process. Biosphere management schemes that do not recognize the underlying socioeconomic realities of a region could significantly conflict with and potentially worsen existing inequities (Solecki 1994). Any proposal should be responsive to the ongoing conservation activities within the region and the interests and understandings of the local populace.

Furthermore, any potential NYMR biosphere reserve must be designed to achieve the three primary functions of biosphere reserves: (1) conservation (restoration in some cases) of natural sites and processes, (2) sustainable development experimentation, and (3) data monitoring and analysis. Each of these elements is intimately linked with the others. The reserve should enable the identification of potential interactions among the three functions (e.g., how sustainable development will promote biodiversity protection).

Operationally, there might be several key programs, campaigns, or foci around which the activities of an NYMR biosphere reserve might be centered. As with most BRs, open space protection and development would likely be a central and initial focus. In the New York urban setting, the goal could be the enhancement of the distribution and connectivity of open spaces. Lessons learned from enhancing open space protection, such as measurement and characterization of long-term benefits and evaluation of the effectiveness of various planning strategies, could provide feedback to assist in the development of the other programs. For example, lessons learned during managing the Hackensack Meadowlands and the Pinelands National Reserves in New Jersey can now be applied to the Highlands Protection Area in the northern part of the state created in 2004.

A New York Metropolitan Region biosphere reserve could be conceived at three different scales as reflective of the varying types of conservation activities already ongoing within the administrative, hydrologic, or metropolitan region. A key geographic question is delineation of the reserve's boundary and its subzones for more specialized planning purposes.

With respect to the outer boundary, BRs most typically have been defined through the use of natural features (e.g., watersheds or relief profiles) and political units (e.g., sets of counties or federal jurisdictions). In terms of composition, BRs either are structured around a central critical conservation core (e.g., Yellowstone Biosphere Reserve) or around a set of smaller, yet still ecologically important, core sites (e.g., Adirondacks–Lake Champlain Biosphere Reserve). Urban biospheres are typically organized in this way, except in cases like Rome where the biosphere planning efforts are centered on the architectural and landscape elements of ancient Rome (e.g., the Coliseum and surrounding area). A physically defined regional BR could be defined for the New York Metropolitan Region. Although the region is centered on the harbor/estuary and the associated local and regional river basins, this condition is not currently reflected in the local political or social culture.

One scenario for a biosphere reserve in the region includes the identification of administratively defined core area around which transition and buffer zones could be defined, similar to that being done in Canberra and Melbourne. The core could include a single area or multiple existing conservation areas in the region. Large and ecological important conservation areas such as Jamaica Bay and the Hackensack Meadowlands have some identity among local and regional managers but have limited, widespread regional recognition. A possible extension of this effort could incorporate activities like those outlined in the Golden Gate Biosphere Reserve in San Francisco and Green Belt Biosphere Reserve in São Paulo examples. These reserves consist of a series of public open spaces extending through the metropolitan zone. Like the San Francisco example, the New York–New Jersey Harbor Estuary is ringed by numerous publicly owned parcels that are managed as parks, recreation areas, and wildlife refuges run by municipal, state, or federal entities. The largest single parcel is the Gateway National Recreation Area, which includes approximately 20,000 acres. The buffer and transition areas of this scenario would include the properties directly adjacent (i.e., bordering) and near (i.e., no more than one kilometer) to the core parcels. Buffer zone analysis developed in the field of conservation biology could be used to determine specific distances for each area.

A second scenario for a NYMR biosphere reserve core extends the first one to include the waters of the harbor estuary area as well as the adjacent conservation land parcels. This scenario, like the first, mirrors the Golden Gate Biosphere Reserve, which includes a large water component consisting of two national marine Sanctuaries. In the New York case, this possible core area could include the New York Upper and Lower bays and adjacent coastal areas (e.g., Hackensack Meadowlands and Jamaica Bay). The buffer and transition zones in this option would be defined by the watershed of the estuary (like the H_2O initiative), and the transition zone would include areas beyond the boundary of the watershed (i.e., approximately a twenty-kilometer-wide zone) that would provide some opportunity to track transboundary environmental impacts such as interbasin water transfers and region-scale, land use-change-derived climatological shifts.

A third biosphere reserve scenario includes the definition of entire metropolitan region as a biosphere reserve. This scenario is motivated by a reconceptualization of the urban biosphere as a tattered yet still intact ecosystem. Satellite images of the region reveal the presence of numerous areas, both large and small, of thriving natural places. On the surface, this regional definition might seem too broad and complex to be useful in potential biosphere reserve planning efforts. To overcome this problem, the county-level scale of government should be the appropriate party to develop this scenario. Throughout the twentieth century, the region was defined via census and regional planning efforts at a county scale. Probably the most recognizable image of the area is the thirty-one-county region defined by the Regional Plan Association (figure 1). Each of these counties has an active parks department that manages protected areas and, importantly, has close and active ties with the citizens, especially those in poorer and less-enfranchised socioeconomic groups.

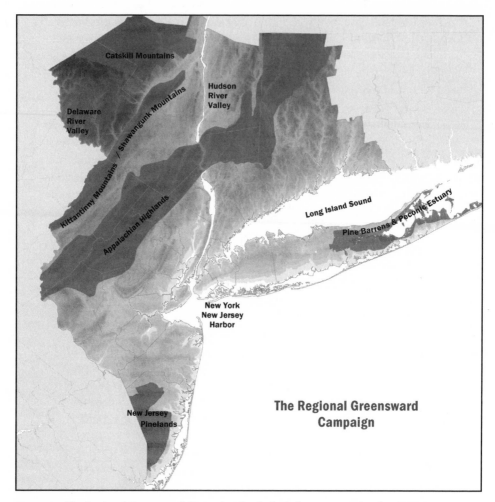

Figure 1 The Regional Greensward Campaign. Dark shaded areas are regional greenspaces proposed for preservation by the New York–based Regional Plan Association. (Map created by Jennifer R. Cox, RPA staff planner, 2006.)

As an example, the Westchester County park system spans more than 17,000 acres in fifty parks and recreational areas. The county government offers a wide range of educational, recreational, entertainment, and sporting events for all citizens. Demonstrating that alliances with higher levels of organization reinforces protected area management goals, the Westchester County Parks Department has the distinction of being the first county in New York State to become accredited by the National Recreation and Parks Association. By extension, a comprehensive biosphere reserve designation weaving together the parks in all thirty-one counties of the NYMR could provide the cohesion, networks, and support needed to transform the region's consciousness and therefore management of its rich though fragmented biosphere.

The principal objective of protecting the core in most biosphere reserves is to preserve critical biodiversity resources. Sustainability is often defined with respect to the level of long-term resource protection. The open space of the New York Metropolitan Region is collectively home to a wide array of species, comparable to similar coastal settings such as the Chesapeake Bay and Delaware Bay. The core of an NYMR reserve will be important with respect to ecological biodiversity and to societal and economic function.

Another important element of this reserve, like other urban BR planning efforts, is that it could become an opportunity for local community groups and other nongovernmental organizations to promote their urban environmental agendas. A New York biosphere reserve could serve as an urban laboratory for sustainability experimentation, such as an urban forestry project for urban heat-island reduction, enhanced use of bicycle commuting along greenways and trails, and the use of conservation areas for flood control. To enhance the societal function of the core for the region as a whole, its ecological function needs to be sustained as well. In turn, to sustain the ecological function of the core, the environmental inputs from the buffer areas (e.g., the surrounding lands directly adjacent to the core areas) and other nearby transition-zone population centers (e.g., within five to twenty kilometers, depending on the BR scenario adopted) need to be better understood as well. Gathering information about these interactions and feedbacks could help foster an increased appreciation of the ecological and environmental interconnections between the various parts of the region among decision makers, stakeholders, and the general public.

Similar to other urban BRs, in the case of the New York reserve, there will be bidirectional connections between the core and adjacent areas. The buffer and transitions areas protect the core, as typically in most BRs, and the core via ecological services and benefits will serve the buffer and transition zones. The protection of the core preserves the ecological integrity of the areas and simultaneously enables it to fulfill its critical role as a social resource of the region.

There are analogous situations in more traditional biosphere reserves where the reserve itself is the center of the social conditions and consciousness of the region; this relationship, however, is rarely described in such explicit terms. For example, what would become of the Greater Yellowstone region without Yellowstone National Park and other protected open space? These lands, like Yellowstone National Park, embody both the ecological and social core of the region. This type of interaction seems to be growing in recognition in São Paulo through the emergence of the Green Belt Biosphere Reserve (São Paulo Forest Institute 2004). Developing this kind of regional consciousness could be the greatest achievement of the New York BR.

Regional planning is challenging in urban areas such as the New York Metropolitan Region where home rule and a splintered political landscape characterize the region (Gunderson, Holling, and Light 1995). Short-term political concerns tend to dominate, and long-term biodiversity and ecological issues are often not understood to have wide-reaching societal effects. Policy responses to biodiversity protection are also hampered by the generally reactive nature of management organizations. Institutional action is often directed at immediate and obvious problems; issues that might emerge fully only after several decades are perceived as less pressing. These issues are compounded by fiscal distress in the region, caused in part by recovery efforts after September 11, 2001, which are focused primarily on rebuilding lower Manhattan.

Several initiatives will help build the necessary foundation for a biosphere reserve strategy: (1) increased communication and cooperation among nongovernmental groups, agencies, and research institutions; (2) methods for defining and entraining potential biodiversity effects into planning decisions, and (3) education and outreach programs. For education programs, a media campaign is needed, with broad dissemination of a carefully written mission statement reflecting the various stakeholder interests involved. Communication and cooperation are developing in the region across a wide range of sectors groups. For example, the H_2O Initiative is encouraging a large-scale watershed approach to understanding the region's ecosystem function (Hiss and Meier 2004). The Nature Network, a coalition of more than three dozen environmental science and education organization of the New York metropolitan area, was launched in April 2005. The goal of this effort is to provide a framework that allows the member organizations to work together to provide the public with a better understanding of the importance of biodiversity and the programs available to protect it.

Designation of a biosphere reserve in the New York metropolitan area could facilitate better understanding of the environmental connections between different parts of the region, be more responsive to potential environmental changes on longer time horizons, and be more flexible in the face of increased environmental uncertainty. By embracing the urban BR strategy, the NYMR could once again serve as a testing ground for new initiatives to meet the environmental challenges known collectively under the rubric of "the transition to sustainability" (Board on Sustainable Development 1999). The goal is for New York City and its environs to be known not only as the "empire city," but also as the "ecological city," a place where the richness of both biological and societal diversity flourishes.

Acknowledgments

We thank Christine Alfsen-Norodom and Benjamin Lane (UNESCO), Roberta Miller (CIESIN/ Columbia University), and other members of the CUBES Urban Biosphere Group for support and stimulating discussion in developing this case study; and Frank Popper (Rutgers University), Rutherford H. Platt (University of Massachusetts Amherst), and Gregory Remaud (NY/NJ Baykeeper) for comments on the paper. We also thank Lauren Sacks (Columbia University) for work on the case study preparation and for help in organizing our Case Study Stakeholders Workshop. We also benefited from key suggestions from Carli Paine and Hugh Hogan.

References

Alfsen-Norodom, C., and B. D. Lane. 2002. Global knowledge networking for site specific strategies: The International Conference on Biodiversity and Society. *Environmental Science and Policy* 5:3–8.

Australian National Sustainability Centre, ANB Working Groups. 2003. Canberra—"Bush Capital" Biosphere Reserve. Available online at http://www.sustainability.org.au/html/content_working_ biosphere.html.

Batisse, M. 1993. The silver jubilee of MAB and its revival. *Environmental Conservation* 202(2): 107–22.

Beatley, T. 1994. *Habitat conservation planning.* Austin: University of Texas Press.

Bennett, M., and D. W. Teague, eds. 1999. *The nature of cities: Ecocriticism and urban environments.* Tucson: University of Arizona Press.

Board on Sustainable Development, Policy Division, National Research Council. 1999. *Our common journey, a transition toward sustainability.* Washington, DC: National Academy Press.

Bonnes, M., ed. 1991. *Urban ecology applied to the city of Rome.* UNESCO MAB Project, Report 4, MAB Italia, Roma.

———, ed. 1993. *Perceptions and evaluations of the urban environment quality: A pluridisciplinary approach in the European context.* MAB-Italia, Enel, Roma.

———. 2000. The "ecosystem approach" to urban settlements: 20 years of the "MAB-Rome Project." Paper presented at the first meeting of the ad hoc workgroup to explore applications of the Biosphere Reserve Concept to Urban Areas and Their Hinterlands—MAB Urban Group. UNESCO, Paris, 9 November.

Bonnes, Mirillia, G. Carrus, F. Fornara, A. Aiello, and M. Bonaiuto. 2003. Inhabitants' perception of urban green areas in the city of Rome: In view of a MAB-Rome Biosphere Reserve. Available online at http://www.unesco.org/mab/urban/ecosyst/urban/doc.shtml.

Bridgewater, P. B. 2002. Biosphere reserves: Special places for people and nature. *Environmental Science and Policy* 5:9–12.

Burrows, E. G., and M. Wallace. 1999. *Gotham: A history of New York City to 1898.* New York: Oxford University Press.

Cortner, H. J., and M. A. Moote. 1999. *The politics of ecosystem management.* Washington, DC: Island Press.

Cranz, G. 1982. *The politics of park design: A history of urban parks in America.* Cambridge, MA: MIT Press.Douglas, I., and J. Fox, eds. 2000. *The changing relationship between cities and biosphere reserves.* Report for Urban Forum of the UK MAB Committee, May.

Frost, P. 2001. Urban biosphere reserves: Re-integrating people with the natural environment. *Town and Country Planning* 70(7–8): 213–15.

Golden Gate Biosphere Reserve Association. 2005. Golden Gate Biosphere Reserve Association. Available online at http://www.nps.gov/ggbr/ggbr.htm.

Gornitz, V., with S. Couch. 2001. Sea level rise and coastal hazards. In *Climate change and a global city: An assessment of the Metropolitan East Coast (MEC) region,* ed. C. Rosenzweig and W. Solecki. Metro East Coast Sector Report of the U.S. National Assessment of Potential Climate Change Impacts. New York: Columbia Earth Institute.

Gunderson, L. H., C. S. Holling, and S. S. Light, eds. 1995. *Barriers and bridges to the renewal of ecosystems and institutions.* New York: Columbia University Press.

Hartig, E. K., V. Gornitz, A. Kolkler, F. Mushacke, and D. Fallon. 2002. Anthropogenic and climate-change impacts on salt marshes of Jamaica Bay, New York City. *Wetlands* 22(1): 71–83.

Haughton, G., and C. Hunter. 1994. *Sustainable cities.* Bristol, PA: Regional Studies Association.

Hiss, T., and C. Meier. 2004. *H₂O—highlands to ocean—a first close look at the outstanding landscapes and waterscapes of the New York/New Jersey metropolitan region.* New York: Geraldine R. Dodge Foundation.

Kates, R. W., and T. J. Wilbanks. 2003. Making the global local: Responding to climate change concerns from the bottom up. *Environment* 45(4): 12–23.

Matsuura, K. 2005. World Environment Day, 5 June, Message from the Director-General of UNESCO. Available online at http://portal.unesco.org/en/ev.php-URL_ID=27747&URL_DO=DO_TOPIC&URL_SECTION=201.html.

Miller, C. 2002. Peninsula proposed as first urban biosphere. *The Age.* Available online at http://www.theage.com.au/articles.

National Research Council, Committee to Review the New York City Watershed Management Strategy, Water Science and Technology Board. 2000. *Watershed management for potable water supply. Assessing the New York City strategy.* Washington, DC: National Academy Press.

New York/New Jersey Harbor Estuary Program. 2002. *Harbor health/human health: An analysis of environmental indicators for the NY/NJ harbor estuary.* New York: Hudson River Foundation and U.S. EPA Region II.

Pires, B. C. C., P. Dale, L. Paolucci, R. Victor, P. M. C. Goncalves, D. Cunha, D., and V. Silveira. 2002. Green belt tourist cluster: A sustainable tourism development strategy in Sao Paulo City Green Belt Biosphere Reserve. Paper presented at the World Ecotourism Summit, Quebec City, Canada, 19–22 May.

Platt, R. H., R. A. Rowntree, and P. C. Muick, eds. 1994. *The ecological city: Preserving and restoring urban biodiversity.* Amherst: University of Massachusetts Press.

Price, M. F., F. MacDonald, and I. Nuttall. 1999. *Review of UK biosphere reserves.* Report to Department of Environment, Transport and the Regions. Oxford: Environmental Change Unit, University of Oxford.

São Paulo Forest Institute, State Department of the Environment. 2004. *Application of the biosphere reserve concept to urban areas, the case of Sao Paulo City Green Belt Biosphere Reserve Brazil.* São Paulo: São Paulo Forest Institute.

Savard, Jean-Pierre L., P. Clergeau, and G. . Mennechez. 2000. Biodiversity concepts and urban ecosystems. *Landscape and Urban Planning* 48:131–42.

Sholtes, W. 2003. Chicago biosphere reserve considered by steering committee. Property Rights Foundation of America Inc. Available online at http://prfamerica.org/ChicagoBioReserveConsidered.html.

Solecki, W. D. 1994. Putting the BR concept into practice: Some evidence of impacts in the rural United States. *Environmental Conservation* 21:242–47.

Solecki, W. D., and F. M. Shelley. 1996. Pollution, political agendas, and policy windows: Environmental policy on the eve of *Silent Spring. Environment and Planning C: Government and Policy* 14:451–68.

Stanvliet, R., J. Jackson, G. Davis, C. DeSwardt, J. Mokhoele, Q. Thom, and B. D. Lane. 2004. The UNESCO biosphere reserve concept as a tool for urban sustainability: The CUBES Cape Town study. *Annals of the New York Academy of Sciences* 1023:80–104.

Tarr, J. A., and R. U. Ayres. 1990. The Hudson-Raritan basin. In *The earth as transformed by human action, global and regional changes in the biosphere over the past 300 years,* ed. B. L. Turner II, W. C. Clark, R. W. Kates, J. F. Richards, J. T. Matthews, and W. B. Meyer, 623–40. New York: Cambridge University Press.

Taylor, R., and J. Hollander. 2003. The new environmentalism and the city-region. Paper presented at the 2003 annual meeting of the American Association of Geographers, New Orleans, March.

UNESCO. 1996. *Biosphere reserves: The Seville strategy and the statutory framework of the world network.* Paris: UNESCO.

———. 1998. Application of the biosphere reserve concept to urban lands and their hinterlands. Paper SC-97CONF. 502/4 UNESCO Advisory Committee for Biosphere Reserves, fifth meeting, July.

———. 2000. *The role of MAB with regard to urban and peri-urban issues.* Paris: UNESCO.

———. 2003. *Frequently asked questions on Biosphere Reserves.* Available online at http://www.unesco .org/mab/brfaq.htm.

United Nations, U.N. Population Division. 1995. *World urbanization prospects: The 1994 revision.* New York: United Nations Press.Waldman, J. 1999. *Heartbeats in the muck: A dramatic look at the history, sea life, and environment of New York harbor.* New York: Lyons Press.

Yaro, R., and T. Hiss. 1996. *A region at risk: The Third Regional Plan for the NY/NJ/CT metropolitan area.* New York: Regional Plan Association.

Zimmerman, R., and M. Cusker. 2001. Institutional decision-making in the New York metropolitan region. In *Climate change and a global city: The metropolitan East Coast regional assessment,* ed. C. Rosenzweig and W. D. Solecki, 149–73. New York: Columbia Earth Institute.

PART THREE

Restoring Urban Nature:
Projects and Process

Part III turns from urban "open spaces" (green or paved, local or regional) to the ecological functions and biodiversity that such spaces may support, with a little human assistance. The opening essay is by plant biologists Steven E. Clemants and Steven N. Handel, collaborators in the Center for Urban Restoration Ecology (CURE), a joint venture of Rutgers University and the Brooklyn Botanic Garden. Their contribution first distinguishes the perspectives of landscape architects and plant ecologists in terms of what makes up a "successful" urban plant community. They then summarize some results from their ongoing program to establish (not "restore") ecological habitats on such barren land features as sanitary landfills. The gigantic Fresh Kills landfill on Staten Island, New York, is the "laboratory" for Handel's students to nurture biodiversity amid a literal landscape of death (Fresh Kills is where the World Trade Center debris was deposited).

Much restoration of plant and wildlife in urban areas is conducted under the rubric of stream restoration. Laurin N. Sievert is a Milwaukee native, a geography graduate student at the University of Massachusetts Amherst, and project manager of the Ecological Cities Project. Her essay is based on her master's thesis research, which examined stream and wetland restoration programs in the Milwaukee River watershed, one of several case studies of urban watershed management conducted by the Ecological Cities Project under a grant from the National Science Foundation.

Industrial brownfields in urban areas are inherently ugly, dangerous, and often ecologically barren. Nevertheless, urban planners, environmental engineers, and natural scientists are collaborating in efforts to restore many such sites to productive human and natural uses. Geographer Christopher A. De Sousa at the University of Wisconsin–Milwaukee summarizes findings of his ongoing research drawing on brownfield remediations in Toronto, Chicago, and Pittsburgh.

Quixotic as some ecological restoration work may seem, potential benefits are not purely numerical, that is, acres replanted, threatened species recovered, salmon returning, or salamanders counted. Andrew Light, geographer and ethicist at the University of Washington, identifies important nonnumerical benefits of ecological restoration, namely the fostering of social contact among people who engage, usually as volunteers, in litter cleanups, clearance of invasive species, and nurturing of more robust biodiversity. Light's concept of "ecological citizenship" also postulates that individuals who engage in such restoration activities gain a strengthened psychic bond to the place and to nature (somewhat akin to Robert L. Ryan's concept of "park adoption" discussed in Part II).

Restoring Urban Ecology

The New York–New Jersey Metropolitan Area Experience

Steven E. Clemants and Steven N. Handel

Interest in restoring urban ecological services and biodiversity is a growing part of modern biology. To protect and restore ecological services in urban areas, two approaches are being tried. *Conservation biology* seeks to keep relatively intact remnants of our plant and animal communities from being destroyed. This conservation tradition dates back about one hundred years and is now a significant academic and public policy pursuit. *Restoration ecology,* a new strategy, seeks to restore and expand ecological services. Restoration aims to restore plant and animal species to areas where they have been eliminated or degraded.

Conservation and restoration share biotic knowledge and theoretical frameworks. Clearly, though, conserving existing biotic conditions at a site is a different matter from attempting to restore the site to some previous "natural" biological state. The latter is a much more difficult task in part because landscapes do not change overnight. Human activities have gradually transformed the landscape and ecological conditions over several centuries in the northeastern United States (including native land use changes) (Cronon 1983), and over millennia in cities of the Old World, as documented by George Perkins Marsh in his seminal 1864 treatise, *Man and Nature* (Marsh 1864/1965), and in recent reviews (Goudie 2000). At what point in this evolving process of change are ecological conditions considered to be "natural"? (For that matter, the natural world itself is also in constant state of evolution.)

Thus, restoration biology pursues a moving target that is very poorly defined for any particular period. Biotic conditions have differed from one time period to another, and our knowledge of biological conditions of any past period is fraught with scientific uncertainty. Furthermore, the present biogeographic context of the site—its physical habitat and biotic milieu—may have changed so radically that "native" species may not be sustainable, and the retention of nonnative ("alien") biological species and communities may be unavoidable and perhaps desirable.

This essay summarizes some examples of restoration efforts that involve the botanical and ecological communities in and around the New York–New Jersey urban complex. Our approach to urban restoration ecology involves applying skills from modern botany and community ecology. The sample pilot studies discussed are teaching us the limits to restoring this historic biodiversity in modified modern urban habitats.

Restoration Ecology versus Landscape Architecture

A caveat is in order concerning the distinction between restoration ecology and landscape architecture, a field that shares some superficial similarities with the former. Landscape architects design and install plant communities and often use native species. The goals of landscape architecture, however, are aesthetic and social, and usually involve management over many years to keep the original landscape design intact. In restoration ecology, the fundamental goals are ecological services and functions, namely the processes and dynamics that are typical of a complex living community. Birth rates and death rates are fundamental to such communities.

In restoration projects, we expect many species to reproduce and spread, even changing their location in the habitat over time. We expect some of the installed species to die out over time because successional forces favor new species. Also, in restoration, we expect plants and animals to be closely associated when determining stable population levels. In these ways, restoration relies more on function than on appearance: after a couple of decades a restored plant and animal community may look very different from the original installation. This outcome would be a success because change is a healthy part of ecological function. By contrast, in most landscape architecture designs, little change is expected or wanted, other than growth of individual plants. Consequently, restoration ecologists study ecological dynamics more than design and construction techniques, and landscape architecture programs rarely include advanced modern ecology. The products of both professions are important and wanted by society, but these products have different settings and goals.

Biodiversity in urban areas provides many benefits (Naeem et al. 1999; Costanza 2001). First, natural habitats serve the social need for a more aesthetic and healthy environment. Second, living plant communities modify the physical world in constructive ways: they clean and moderate the microclimate, promote groundwater infiltration, retard flooding and soil erosion, and provide habitat for wildlife (Daily 1997). Third, living plant communities enhance property values in locales where people wish to live near greenspaces (Daily 1997). Restoration activities, however, must address many challenges to creating historic and self-sustaining natural habitats. In urban areas, for example, the extensive infrastructure, homes, roads, industrial centers, and shopping areas fragment the landscape into small, oddly shaped patches. Unlike the dimensionless earlier theories of ecology, contemporary urban ecology focuses on such spatial constraints to understand what is feasible in reestablishing biodiversity.

The Hackensack Meadowlands (New Jersey)

In one of our first studies, on a landfill in the New Jersey Meadowlands (Robinson and Handel 2000), we are trying to bring back many tree and shrub species to a derelict landscape covered only with alien weeds. Although this landfill had been left alone by the responsible municipality, the Town of Kearny, New Jersey, for almost twenty years, no early successional or native species were found there. Surrounded by highways, dense urban communities, railroad yards, and saline marshes, this landfill was isolated from sources of native plants and animal species. Birds, which serve as agents of species introduction, had no reason to visit the barren site: there were no nesting areas, no perches, and no food. On all sides of this landfill were paved and hot surfaces that deterred the appearance of new species on the landfill. Vegetation in nearby areas was primarily alien and invasive species, predominantly mugwort *(Artemisia vulgaris)* and phragmites *(Phragmites australis)*, rather than those associated with natural early successional habitats in this region. Finally, the soil structure and horizons so important for healthy plant communities were lacking. The engineers who designed and closed the landfill did not have ecology in mind; their only goal was to concentrate and cover solid waste.

In urban areas, deficiencies in plant community are matched by peculiar and incomplete animal communities (McKinney 2002). Many of the large predators, wolves and large felines, originally found in the New York–New Jersey regions are long gone. Suburbia favors large deer populations, which destroy many plants and plant communities that restoration ecologists seek to nurture (Waller and Alverson 1997). The interplay of animals and plants is a critical part of modern restoration ecology.

Apart from the familiar problem of rampant deer herbivory, other less obvious plant-animal interactions are also important (Handel 1997). Mutualisms between plants and animals are critical for sustainable and healthy natural communities. Even healthy plants cannot reproduce and species populations cannot grow unless pollinators and seed dispersers, which in this region are usually animals, are present. We are just learning how to bring back populations of these animals as partners to the plants.

Although many plant species are found in commercial nurseries, most animals that are needed for plant reproduction are not commercially available and must be attracted to a restored plant community from the surrounding region. In urban area, that is a great challenge, and many animal species may never be encountered. Living species in the soil are similarly important (Allen 1991). Invertebrates and fungi are necessary for long-term biotic health but also cannot be obtained from commercial sources. Some mycorrhizae fungi, on plant roots that facilitate nutrient and water uptake, are available from suppliers, but the full complement of necessary species required natural functioning are not (Dighton 2003).

Experimental Restorations at Fresh Kills Landfill

Our first test case of restoration in the New Jersey Meadowlands suggested the need for a much more comprehensive experimental approach to urban restoration. With the support of the City of New York Department of Sanitation and the National Science Foundation, we have attempted a wide series of experiments at the Fresh Kills landfill in Staten Island, the largest landfill in North America, covering almost 1,100 hectares. After being closed in 1999, Fresh Kills was reopened in 2002 to receive debris removed from the destruction of the World Trade Center on September 11, 2001. It is now closed again, awaiting its transformation into a public parkland.

Closed landfills are usually capped by clays or heavy plastic, which is then covered by a layer of clean soil fill to protect the barrier layer against damage from sun and precipitation. The clean soil, obtained from local sources, is stabilized by hydroseeding a dense cover of fast-growing perennial grasses. The design goals are to protect the solid waste from being exposed and the protection of groundwater from chemicals leaching down from the landfill; no biotic or natural habitat goal is reflected in the engineering design. To realize the potential for Fresh Kills someday to serve as a huge urban park or natural refuge, engineering design must be tempered with a strong contribution from restoration ecology.

Project Design

For about a decade, we have conducted a multifaceted experiment in urban restoration ecology at Fresh Kills. The original salt marsh is now covered by almost sixty meters of solid waste, so recreation of the historic marsh community is impossible. A reasonable goal for restoration would be meadows and woodlands, typical of the coastal plain of New York. We are accumulating records of exactly which species used to grow on the upland coastal plain. Details of this data set, *New York Metropolitan Flora* (Moore et al. 2002), are mentioned below. An experiment testing the performance of native trees and shrubs was installed on the site to learn which species can grow on the shallow engineered soil.

We also want to learn whether the scale or intensity of planting affects ecological function and long-term success. Four sizes of woody plant patches were installed at the site. Each patch type had seven, twenty-one, forty-two, or seventy woody plants installed. Only seven species were used, so the patches represented one, three, six, or ten individuals in each species. Each patch size was replicated five times, and all twenty patches were planted on a slope of this landfill (figure 1).

An economic issue parallels the ecological question: because landfill managers

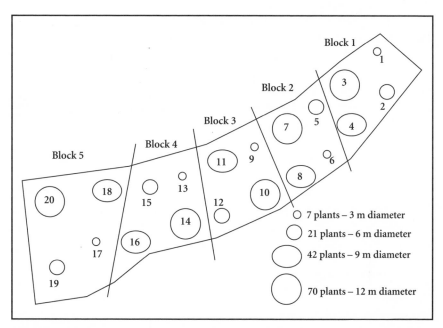

Block 1

Block 2

Block 3

Block 4

Block 5

○ 7 plants – 3 m diameter

○ 21 plants – 6 m diameter

○ 42 plants – 9 m diameter

○ 70 plants – 12 m diameter

Figure 1 Arrangement of experimental patches of woody plants at Fresh Kills landfill, Staten Island, New York. Twenty patches of seven species vary in size, containing seven, twenty-one, forty-two, or seventy plants. These patches test how scale of restoration planting may change ecological functioning.

have limited funds, we are investigating whether small patches of plants would survive and spread across the large landscape spontaneously. In other words, is it more cost effective to plant many small, scattered patches or fewer but much larger patches of plants for long-term success? This coalescence of ecological and economic questions can be critical for future restoration ecology research.

Several preliminary conclusions can already be reported from the Fresh Kills experiment. Although it was a very dry and physically stressed site, the plants chosen grew relatively well on the landfill slope. Many of the patches are much larger than at installation, and individual plants have grown and are reproducing. Clonal growth, the vegetative spread of individual plants, in contrast to seedling additions, has characterized plant and patch growth. The native roses and sumacs in the patches now have many stems and cover many square meters. The plants have produced larger clusters that offer better habitat for both vertebrates and invertebrates. The patches are also slowly changing and improving the soil beneath the plants. Each year, leaves and woody litter from the plants scatter on the ground and decay, adding to the organic matter in the final soil cover. This process enriches the site and facilitates survival and growth of these plants into the future.

We have learned that the larger patches accumulate relatively more litter. The many stems in the large patches act as traps, preventing wind from blowing away dead leaves. In smaller patches where many plants are near the patch edges, dead leaves scatter across the site and away from the installed plant individuals. In addition, the larger patches have developed a deep shade like a natural thicket, and the original hydroseeded grass cover is dying, which is desirable because the heavy grass cover impeded germination of seeds and growth of new seedlings of woody plants. This negative interaction has been seen for many years on mining sites where reforestation is the preferred end use (Burger and Torbert 1992, 1999). Heavy grass cover kills woody plant seedlings directly by shading and competition for space, and indirectly by harboring large populations of rodents that eat woody seedlings as they appear. The developing shade of the large patches kills grass, which facilitates the opportunity for new woody species and individuals to succeed.

Seed Propagation and Pollination

Growth and flowering of our originally installed plants is only one part of the demographic process of restoration. All natural communities interplay with the surrounding landscape. On Staten Island—a densely populated and industrial area—few remnants of nonurban landscapes are left. Would seeds from additional species ever be carried into this large landfill? We tested this premise by placing seed traps under many of the trees in the twenty patches, and we also placed some traps in the open grassland. We found that birds, even on densely populated Staten Island, would bring in thousands of seeds of native woody species (Robinson, Handel, and Mattei 2002).

These extraordinary results suggested that this critical link in nature could be reestablished despite even the most stringent landscape conditions. Seeds of more than twenty new plant species were added to our site in each year of the first three years of study. The number of seeds, however, changed from year to year. In dry years, fewer seeds were available and spread to our site. In another year, a large part of an adjacent small woodland was cut down for commercial development. This habitat destruction was also correlated with few seeds coming into our site, *suggesting that even small urban and suburban woodlots play a very important role in the future restoration of healthy habitats.* In addition to numbers of seeds, urban remnant habitats represent plant survivors of urban stresses. Seeds from these remnants may represent genotypes that can best succeed in today's stressful urban physical conditions (Handel, Robinson, and Beattie 1994).

The growth of our planted individuals was encouraging, as was the addition of other seeds and species from surrounding habitat remnants. Self-sustainability of our plantings, however, must mean that the installed individuals themselves make seed. For many native trees and shrubs, native pollinators must visit flowers on these plants (Handel 1997). Would native bees visit these relatively few plants sur-rounded by hundreds of hectares of grasslands and urban infrastructure?

To address this question, a study was conducted involving bee species at the ex-perimental planting in comparison with bees visiting the same plant species in urban parks surrounding Fresh Kills (Yurlina and Handel 1995; Yurlina 1998). This study produced very optimistic findings. More than seventy native bees spe-cies were found on flowers at Fresh Kills landfill. This number was similar to that of bee species found in old native habitats on Staten Island. More critically, the number of flowers on the landfill planting that set seed was statistically the same as the percent of flowers on same plant species in natural parks that set seed. This finding suggests that this link in nature–pollinators invading a large restored site and facilitating seed set–can occur even in the largest city in the United States.

Germination of Seeds

Finally, we tested whether seeds on the ground would in fact germinate and emerge, starting large populations of woody plants (Robinson, Handel, and Mattei 2002). We planted thousands of seeds of twenty-seven native species in another part of Fresh Kills landfill and followed their fate for three years. Very few of the seeds succeeded. The poor soil conditions and the competition from the dense fescue grasses challenged reproductive success. For restoration on landfills, soil quality, competition from grasses used for erosion control, and fate of introduced seeds form a trio that cannot be separated. Success of new species invading the sites re-quires microsites in the soil and a lack of competition from plants planted solely for engineering needs.

The Long View

A final requirement of restoration on urban engineered sites is adequate time for biotic success. We define it in three ways.

First, time is needed for more native plant species to reach to restored patches, find a microsite to begin their growth, and reach reproductive age before being eliminated by enemies such as herbivores and diseases. Virtually all natural com-munities through successional time change. Restorations in urban areas need sig-nificant time for the slow processes to occur. Very often, restoration project contracts are written that are only monitored for three years before success is mea-sured and rated. A modification of this usual construction procedure will be neces-sary to make our urban restorations truly successful.

Second, our studies have shown that there is a long-term need for management of the projects. Invasive species often come in from the surrounding areas and can destroy the biodiversity we wish encouraged (Mack et al. 2000). Depending on the quality of the surrounding habitat remnants, these invasions may come in quickly or slowly. For example, in a forest fragment north of Philadelphia, alien vines destroyed much of the native forest and crippled attempts to restore new native trees to this preserve (Robertson, Robertson, and Tague 1994). Some labor is needed to destroy small populations of invasive species before they overwhelm the native species we wish to encourage (Sauer 1998).

Third, mature communities often include many more species than those found in early succession areas. To bring back species that require deep, rich soil or heavy shade for survival, restoration plantings may have to be added several years after the initial site treatment. Thus, there must be an *administrative organization* in place that remembers the original goals of the restoration and has the administrative ability to return to the site several years later. Funds must be reserved for this later stage of restoration. Many urban land managers do not have a long-term perspective. In fact, fiscal needs are often defined for only short time periods. Ecological restoration needs time for an organization to work on one site. Some private organizations that run urban parks have the institutional memory to keep working on a site for decades (e.g., Toth 1991). Civic organizations using public funds must design institutional methods to supply the time needed for realistic results.

New York Metropolitan Flora: Patterns of Urban Biodiversity

New York Metropolitan Flora is a project of the Brooklyn Botanic Garden (Moore et al. 2003). The plant occurrence database amassed by this project represents an important record of the local environment over the past 150 years. We have been working with these data to determine how we can analyze them and what the changes in range of various species mean in terms of the urban environment and its changes. (A more extensive account of the origin of the data, biases in the data, and the statistical analysis is presented in Clemants and Moore 2005.)

The data used here come from the New York Metropolitan Flora database, AILANTHUS (figure 2). We currently have more than two hundred thousand records of plant occurrences in the metropolitan region. These data have come from a variety of sources but particularly from herbarium specimens housed at eleven herbaria of the Northeast, extensive published and unpublished lists from literature, and five years of field work in the region. The woody plant data are represented by nearly one hundred thousand nonduplicated records and at least one hundred records per year for each year in the past century.

Two characteristics of the data are *distribution* and *change in range* (change index) over the past century. The distribution of species often indicates which environmental parameter might be most important restricting the range of a species. For instance, staggerbush (*Lyonia mariana*) (figure 3) is nearly restricted to the coastal plain of New York and New Jersey, which suggests that the soils or other characteristic of this physiographic province are critical to the limits of its range.

The change index is calculated using the methods presented in Telfer. Preston,

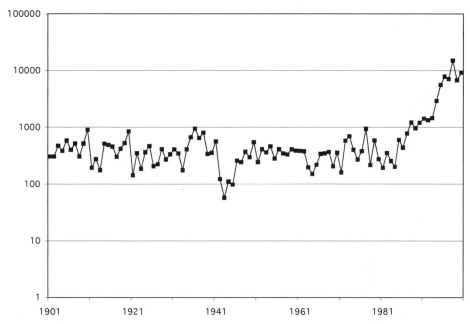

Figure 2   Number of records of plant species' occurrences per year in the database. Note that this graph is a log scale. Over the century, the number of records per year was less than two hundred per year, but there is a flurry of intense data collection since 1990, when NYMR was supported for intense fieldwork.

and Rothery (2002). The basic idea is to select two periods—in our case 1901–50 and 1951–2000—with available comparable data blocks for both periods. Counting the number of blocks in which a species occurs for both periods and graphing the early period's species counts against the later period will give an average change in number of blocks, which would represent the changes in sampling density. The divergence of a species from this average can therefore be attributed to *changes in the range of the species* (plus error). The magnitude of this divergence represents the magnitude of range change over the century.

The change index represented the direction and magnitude of this difference (see Telfer, Preston, and Rothery 2002 for actual calculation). For instance, *Celastrus orbiculata*, oriental bittersweet, is a highly invasive species introduced into the New York area in the early 1900s that spread rapidly after the middle of the twentieth century. Its change index is +3.34, the highest for any species studied. On the other hand, the related native species, *Celastrus scandens*, American bittersweet, has apparently declined, with a change index of –1.05. These data can now be used to examine some of the characteristics of the urban environment and how they affect various plant species.

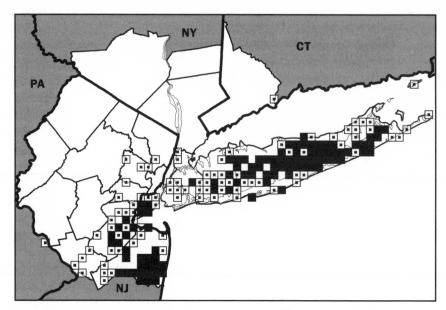

Figure 3 Distribution of staggerbush, *Lyonia mariana*. Filled squares indicate the species was found within the square in the past twenty-five years. An open square with a dot in the middle indicates that the species was found within the square before 1980 but not seen since.

Figure 4 Kudzu distribution. Black squares indicate the species was found within the square in the past twenty-five years. An open square with a dot in the middle indicates the species was found within the square before 1980 but not seen since.

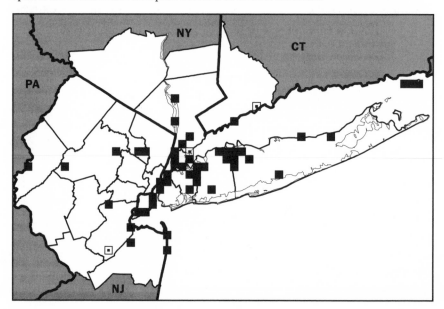

It is well known that the *urban climate* is distinct from surrounding rural areas. Cities are warmer, particularly in and near their downtowns, a condition known as the "urban heat island" (Pickett et al. 2001). One species that flourishes in warmer climates is Kudzu (*Pueraria lobata*), which has been predominately collected in the denser urban areas (figure 4). Kudzu is a short-day plant, becoming reproductive only when nights are relatively long and blooming very late in the fall. Under normal climatic conditions in the New York region it would rarely set seed. Under the heat island, however, frosts are delayed and the plant will make seed more frequently.

Equally well known is the effect of *urbanization on soil*. Particularly apparent are reduced soil organisms and the heightened alkalinity as water drains off of concrete pavements and other hard surfaces. These two effects have markedly reduced the populations of acid-loving mycorrhizal members of the Ericaceae, the heath family. Table 1 shows all woody members of the Ericaceae and their corresponding change index.

The change index data can also illustrate the risk posed by *invasive species* and which species have shown the greatest increase in range. The average change index for the 47 nonnative species is +0.75, whereas the average for the 215 native species is –0.16. These numbers suggest that nonnative species became much more abundant over the past century and that native species are in general slightly declining during the same period.

Examining the species with the highest and lowest change index scores shows a similar trend. Table 2 gives the top-ten scoring species; only one species is native. Table 3 gives the lowest-scoring ten, and only one species is nonnative.

Developing Restoration Goals

These efforts to restore small native communities are grounded on the assumption that we know what we want the biodiversity to be in the future area. A restoration team cannot design appropriate plant communities that can nurture native animals unless it knows with some accuracy what was there in the past. A critical foundation for all ecological work is the accurate floristic record of which plant species were once present in the landscape (Egan and Howell 2001). Partnership between modern botanists and restoration ecologists must occur before a spade is put into the ground. In the New York–New Jersey region, detailed botanical research has occurred for over a century, and the results of this work are being collated for an understanding of past biodiversity. These data are also critical to understanding what is feasible and practical to restore in our urban habitats.

A new academic and practical approach to enhancing urban biodiversity is emerging. One organization that seeks to promote this synergy is the Center for Urban Restoration Ecology (CURE), a joint project of the Brooklyn Botanical

Table 1   Change Index Values for Ericaceae (Heath family) Species

Andromeda glaucophylla	−1.80
Arctostaphylos uva-ursi	−1.35
Chamaedaphne calyculata	−0.61
Chimaphila maculata	−0.61
Chimaphila umbellata	−2.07
Epigaea repens	−0.85
Gaultheria procumbens	−0.45
Gaylussacia baccata	−0.39
Gaylussacia dumosa	−1.96
Gaylussacia frondosa	−0.66
Kalmia angustifolia	−0.43
Kalmia latifolia	−0.08
Kalmia polifolia	−0.20
Leucothoe racemosa	−0.54
Lyonia ligustrina	−0.60
Lyonia mariana	−0.54
Rhododendron canadense	−1.73
Rhododendron maximum	+0.13
Rhododendron periclymenoides	−0.39
Rhododendron prinophyllum	−1.45
Rhododendron viscosum	−0.46
Vaccinium angustifolium	−0.83
Vaccinium corymbosum	−0.55
Vaccinium macrocarpon	−1.06
Vaccinium oxycoccos	−0.74
Vaccinium pallidum	−0.14
Vaccinium stamineum	−0.17

Table 2   The Top Change Index Scores, Indicating a Growth in Range

Celastrus orbiculata	3.34
Lonicera morrowii	2.79
Rosa multiflora	2.79
Elaeagnus umbellata	2.47
Ampelopsis brevipedunculata	2.34
Morus alba	2.30
Acer negundo	1.92
Ailanthus altissima	1.90
Rhamnus frangula	1.76
Berberis thunbergii	1.75

Note: Only *Acer negundo* is a native species.

Table 3 The Lowest-Change Index Scores, Indicating
a Contraction of Range

Rubus argutus	−2.62
Rubus canadensis	−2.62
Salix serissima	−2.62
Rubus setosus	−2.51
Rhamnus alnifolia	−2.38
Lonicera dioica	−2.23
Crataegus uniflora	−2.10
Symphoricarpos albus	−2.09
Salix pentandra	−2.09
Chimaphila umbellata	−2.07

Note: All except *Salix pentandra* are native species.

Garden and Rutgers University (jointly administered by the authors of this essay). CURE has four broad goals:

1. To understanding patterns of urban biodiversity
2. To provide protocols for successfully restoration projects
3. To encourage urban restoration
4. To train students and professionals in urban restoration.

Some outgrowths of this collaboration have been the New York Metropolitan Flora Project, ongoing research at the Fresh Kills landfill, lectures, press releases, and demonstration projects. Environmental education is teaching the public, especially schoolchildren, that ecological services are needed in urban areas where most citizens live. As academic restorationists begin to collaborate with governmental entities to improve environmental health, progress in restoration ecology may become more rapid and noticeable.

References

Allen, M. F. 1991. *The ecology of mycorrhizae.* Cambridge, UK: Cambridge University Press.

Burger, J. A., and J. L. Torbert. 1992. *Restoring forests on surface-mined land.* Virginia Cooperative Extension Publication 460–123.

———. 1999. Status of reforestation technology: The Appalachian region. In *Proceedings of enhancement of reforestation at surface coal mines: Technical interaction forum,* ed. K. C. Vories and D. Throgmorton, 95–123. Alton, IL: USDI Office of Surface Mining: and Carbondale, IL: Coal Research Center, Southern Illinois University.

Clemants, S. E., and G. Moore. 2005. The changing flora of the New York Metropolitan Region. *Urban Habitats* 3:192–210. Available online at www.urbanhabitats.org.

Costanza, R. 2001. Visions, values, valuation, and the need for an ecological economics. *BioScience* 51:459–68.

Cronon, E. 1983. *Changes in the land: Indians, colonists, and the ecology of New England.* New York: Hill and Wang.

Daily, G. C., ed. 1997. *Nature's services: Societal dependence on natural ecosystems.* Washington, DC: Island Press.

Dighton, J. 2003. *Fungi in ecosystem processes.* New York: Marcel Dekker.

Egan, D., and E. Howell, eds. 2001. *The historical ecology handbook: A restorationist's guide to reference ecosystems.* Washington, DC: Island Press.

Goudie, A. 2000. *The human impact on the natural environment.* 5th ed. Cambridge, MA: MIT Press.

Handel, S. N. 1997. The role of plant-animal mutualisms in the design and restoration of natural communities. In *Restoration ecology and sustainable development,* ed. K. M. Urbanska, N. R. Webb, and P. J. Edwards, 111–32. Cambridge, UK: Cambridge University Press.

Handel, S. N., G. R. Robinson, and A. J. Beattie. 1994. Biodiversity resources for restoration ecology. *Restoration Ecology* 2:230–41.

Mack, R. N., D. Simberloff, W. M. Lonsdale, H. Evans, M. Clout, and F. Bazzaz. 2000. Biotic invasions: Causes, epidemiology, global consequences and control. *Issues in Ecology, no. 5.* Washington, DC: Ecological Society of America.

Marsh, G. P. 1864/1965. Man and nature: Or, physical geography as modified human action. Cambridge, MA: Harvard University Press.

Moore, G., A. Steward, S. Clemants, S. Glenn, and J. Ma. 2003. *An overview of the New York Metropolitan Flora Project. Urban Habitats.* Available online at http://www.urbanhabitats.org/v01n01/nymf_pdf.pd.

McKinney, M. L. 2002. Urbanization, biodiversity, and conservation. *BioScience* 52:883–90.

Naeem, S., F. S. Chapin, R. Costanza, et al. 1999. Biodiversity and ecosystem functioning: maintaining natural life support processes. *Issues in Ecology, no. 4.* Washington, DC: Ecological Society of America.

Pickett, S. T. A., M. L. Cadenasso, J. M. Grove, C. H. Nilon, R. V. Pouyat, W. C. Zipperer, and R. Costanza. 2001. Urban ecological systems: Linking terrestrial ecological, physical, and socioeconomic components of metropolitan areas. *Annual Review of Ecology and Systematics* 32:127–57.

Robertson, D. J., M. C. Robertson, and T. Tague. 1994. Colonization dynamics of four exotic plants in a northern Piedmont natural area. *Bulletin of the Torrey Botanical Club* 121:107–18.

Robinson, G. R., and S. N. Handel. 2000. Directing spatial patterns of recruitment during an experimental urban woodland reclamation. *Ecological Applications* 10:174–88.

Robinson, G. R., S. N. Handel, and J. Mattei. 2002. Experimental techniques for evaluating the success of restoration projects. *Korean Journal of Ecology* 25(1): 1–7.

Sauer, L. J. 1998. *The one and future forest: A guide to forest restoration strategies.* Washington, DC: Island Press.

Telfer, M. G., C. D. Preston, and P. Rothery. 2002. A general method for measuring relative change in range size from biological atlas data. *Biological Conservation* 107:99–109.

Toth, E. 1991. *An ecosystem approach to woodland management: The case of Prospect Park.* National Association for Olmstead Parks Workbook Series, vol. 2, technical notes pp. 1–11.

Waller, D. M., and W. S. Alverson. 1997. The white-tailed deer: A keystone herbivore. *Wildlife Society Bulletin* 25(2): 217–26.

Yurlina, M. E. 1998. Bee mutualisms and plant reproduction in urban woodland restorations. Ph.D. thesis, Rutgers University, New Brunswick, NJ.

Yurlina, M. E., and S. N. Handel. 1995. Pollinator activity at an experimental restoration and an adjacent woodland: Effect of distance. *Bulletin of the Ecological Society of America* 76(2, suppl.): 292.

Urban Watershed Management

The Milwaukee River Experience

Laurin N. Sievert

A National Resource Council study, *New Strategies for America's Watersheds*, reports, "Successful watershed management strives for a better balance between ecosystem and watershed integrity and provision of human social and economic goals" (NRC 1999, 270). That is, contemporary urban watershed management must recognize and achieve balance between multiple goals, strategies, and interests, including those of both people and nature.

To achieve these ends, new approaches to watershed management necessitate innovative partnerships and collaborations among scientists, resource practitioners, and public interest groups. Further, basinwide management strategies are needed to manage watersheds as systems and to optimize geographic distribution and connectivity of ecological restoration projects (Franklin 1992). In cities, restoration, rediscovery, and celebration of waterways can be effective in reuniting urban neighborhoods (Rome 2001).

Although the degradation of many urban watersheds in the United States is well documented, there have been fewer studies investigating their recovery. Furthermore, although antidotal evidence indicates that some urban watersheds are improving as a result of coordinated watershed management, more research is needed to identify and document new approaches to managing these systems. This task is complicated by a lack of consistent data at the national level documenting the physical, chemical, and biological status of our water resources (NRC 1992).

This essay summarizes a recent study of innovative approaches to upgrading the Milwaukee River basin in southeastern Wisconsin at multiple scales. This appropriate mix of management strategies and objectives is helping improve water quality and ecosystem health while promoting a greater sense of community in the medium-sized watershed.

To assess the various public and private programs designed to protect and restore watershed health in the Milwaukee River basin, a survey of the root causes of degradation of water resources and synthesis of available data for recent regulatory and management programs, grassroots initiatives, and academic research was undertaken. Throughout the study, both ecological function and the development of a greater sense of community are considered. It is hoped that recent experience in the Milwaukee River basin will inspire and inform comparable efforts elsewhere.

The Milwaukee River Watershed

The Milwaukee River basin in southeastern Wisconsin consists of a network of four adjoining waterways: the Milwaukee, Menomonee, and Kinnickinnic rivers and Cedar Creek. Owing to the basin's size and drainage pattern, it is further divided into six subwatersheds. In sum, the basin covers a land area more than 850 square miles in size with more than six hundred miles of perennial streams, eighty-seven lakes and ponds larger than five acres in size, and thirty-five miles of shoreline along Lake Michigan.

The basin's landscape is diverse (figure 1). Its northern headwaters are largely undeveloped and protected as part of the Kettle Moraine State Forest, and the western portion of the basin is an amalgamation of suburban development and agricultural lands. In contrast, the southern basin is almost entirely metropolitan, with more than one million residents (WDNR 2001a).

The basin is also complex politically. Laying within portions of seven counties in southeastern Wisconsin (Dodge, Fond du Lac, Milwaukee, Ozaukee, Sheboygan, Washington, and Waukesha counties), it encompasses part or all of thirteen cities and twenty-four villages. All surface waters from these communities ultimately discharge into Lake Michigan at Milwaukee's downtown harbor.

Since 1970, the population of the Milwaukee River basin has increased by only 2.2 percent, although this population change is unevenly distributed. At the same time, the City of Milwaukee has experienced a decline in absolute population as residents have sprawled into adjoining suburbs south, west, and north of the central city. Hence, population in nearby counties has changed dramatically, and urban sprawl reaches into once rural hinterlands. For example, whereas in 1970 Milwaukee County accounted for 82 percent of the basin's total population, today it accounts for only 74 percent. In contrast, nearby Washington, Ozaukee, Fond du Lac, Sheboygan, and Waukesha counties grew by 89, 64, 25, 24 and19 percent, respectively, over the same period (WDNR 2001b).

Management Issues and Stakeholders

Because of its natural and population structure, the Milwaukee River basin faces a wide variety of water quality and quantity problems typical of suburbanizing watersheds in the United States. Water quality concerns in the Milwaukee River basin include point and nonpoint pollution, habitat degradation, and diminished recreational opportunities. Of these issues, combined sewer overflows and public beach closings are the most controversial. Water quantity concerns include flooding and groundwater depletion connected to regional drinking water supply issues.

To confront these issues, watershed organizations at various scales have joined

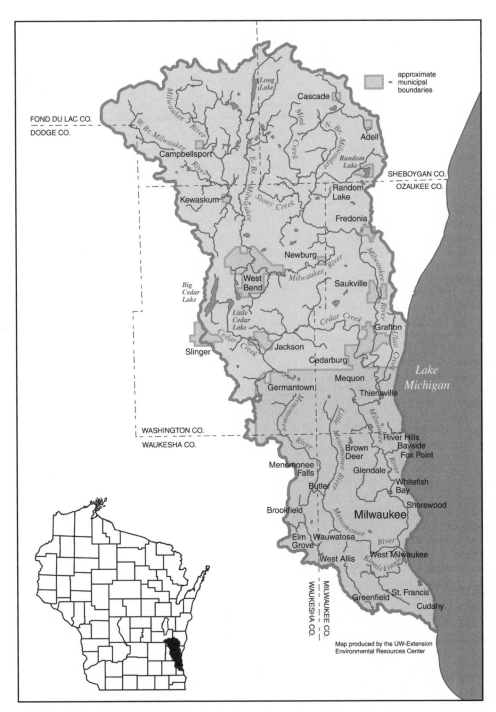

Figure 1 Map of the Milwaukee River Basin. (From University of Wisconsin–Extension Environmental Resources Center).

forces and focused their attentions away from individual problems and are think-ing more holistically about the watershed. This change has occurred at federal, state, and local levels of government as well as in the private and nonprofit sectors.

At the federal level, the U.S. Environmental Protection Agency (EPA) began re-focusing its efforts through the implementation of section 303(d) of the federal Clean Water Act. This legislation requires states to identify polluted waters that do not meet specific water quality standards for inclusion on the EPA's list of impaired waters. For each site listed, states must establish a comprehensive cleanup plan specifying a total maximum daily load (TMDL), which determines the amount by which all sources of pollution need to be reduced to meet the state water quality standards. Although TMDLs must account for both point and nonpoint sources of pollution, implementing them in watershed management increasingly requires focusing efforts on reducing nonpoint sources of pollution, such as nutrients, bac-teria, and sediments that are typically transported in urban and agricultural runoff. In addition, TMDLs have increased attention placed on other factors affecting water quality, such as stream channel alteration, habitat degradation, and other physical modifications to the watershed. Point sources of pollution are largely regulated through the National Pollutant Discharge and Elimination System per-mitting process (U.S. EPA 2004).

In the Milwaukee River basin, eighteen impaired water bodies have been identi-fied and included on Wisconsin's 303(d) list. The majority of pollutant sources for the river segments listed are associated with urban land uses, including bacteria, wetland loss, and sediments. Of the water bodies identified in the watershed, no TMDL plans have been established (WDNR 2001b).

Many problems have been identified with the TMDL approach to watershed management. To begin, the EPA is ill equipped to process and evaluate the number of proposals it receives for inclusion on its 303(d) list. Many states do not have ad-equate water quality data to evaluate the status of many water bodies within their boundaries. Moreover, funding is not available to assist states in developing TMDL implementation plans.

In 1995, the Wisconsin Department of Natural Resources (WDNR) began man-aging its land and water resources by twenty-three geographic map units (GMUs) according to major drainage basins. GMUs, more commonly referred to as "ba-sins," emphasize the natural boundaries, structure, function, and interconnected-ness of land and water resources (WDNR 2001a). This organizational restructuring reflected a significant shift in state and federal policy toward implementing new "eco-region" approaches to resource management.

In each of Wisconsin's GMUs, local partnerships involving a variety of govern-mental and nongovernmental stakeholders have been established. These partner-ships serve in an advisory role to the WDNR and foster local work groups and improved communication between all interests and activities in a basin. Ultimately, the goal of WDNR's basin initiative is to facilitate more citizen-driven, participatory, decision-making processes in land and water resource policy.

Many watershed organizations are looking toward public-private partnerships to solidify goals and work cooperatively to address watershed goals, set priorities, and initiate projects. Drawing on the knowledge and resources of multiple organizations, partnerships allow for broader visions and a larger network of ideas.

To this end, the WDNR and the University of Wisconsin–Extension initiated the Milwaukee River Basin Land and Water Partners Team in 1998. Members of the partnership include businesses, nonprofit groups, public agencies, educational institutions, organizations, and individuals sharing an interest in the environmental and economic health of the Milwaukee River basin. Their initiatives are comprehensive and include research and project implementation, environmental education, and public policy recommendations.

New Directions in Watershed Management

New watershed approaches may be contrasted with earlier, more "traditional" watershed approaches, where watershed management was largely defined by many fragmented structural projects initiated by centralized governmental authorities. This approach is often referred to as a top-down approach, as direction was given by an overreaching government agency. In addition, this approach traditionally addressed only a single problem at a time, such as flooding.

New approaches to watershed management are more "organic" in nature. Characteristics of the new approach include a decentralized structure of governmental and nongovernmental stakeholders sharing in decision making. In addition, the new approach involves creative partnerships to establish and oversee common goals, share resources, name priorities, and exchange information. The goals of watershed management continually evolve to address issues that were largely ignored in the past, such as public participation, environmental education, and environmental justice (Born and Genskow 2001).

Although it is difficult to articulate a "one size fits all" definition of new watershed management approaches, researchers at the University of Wisconsin–Madison have found that such approaches generally share the following characteristics (Born and Genskow 2001):

1. Organize by watersheds and subwatersheds as their primary analytical and management units
2. Address a broad spectrum of issues
3. Exhibit a systems orientation
4. Incorporate multiple means and include goals pertaining to healthy ecosystems, economic returns, and resource management

5. Assess decision-making processes based on a combination of biophysical-science, social, and economic factors as well as local knowledge
6. Include interactions among multiple agencies and multiple levels of government
7. Emphasize influential and voluntary participation of multiple local and nongovernmental interests
8. Demonstrate collaborative, problem-solving, planning, and management orientations.

Because of the organic nature of this broader type of watershed management approach, evaluating its effectiveness presents new challenges. Moreover, whereas the ultimate goal of coordinated watershed management may be to achieve a measurable environmental outcome, the nature and breadth of the new "systems approach" requires a combination of both quantitative and qualitative indicators of progress. Finally, at different stages of this evolving management system, various indicators may become more or less relevant (Born and Genskow 2001).

Despite the challenges, many innovative projects throughout southeastern Wisconsin are contributing to an ecologically, economically, aesthetically, and socially enhanced Milwaukee River basin. The remainder of this essay examines a few of these projects initiated by a range of stakeholders to demonstrate the various levels of complexity in which watershed issues are being addressed throughout the basin. Many of these examples apply to the urbanized downstream portions of the watershed where population densities are highest and modification of the natural environment has been most pronounced.

Toward Collaborative Watershed Management

Resource managers and their partners are improving the Milwaukee River basin through a variety of efforts. These initiatives address both ecological and economic needs of the communities they benefit.

Economic Opportunities

In the 1930s, to protect citizens against flood losses, the Milwaukee County Parks Commission adopted a river parkway system recommended by Frederick Law Olmsted. This early foresight left Milwaukee with a rich legacy of parks, and public access to the waterfront in downtown Milwaukee that remains today (Riley 1998, 13).

In keeping with its responsibility to protect navigable waters and public commons according to the Wisconsin Public Trust Doctrine and to create better public access to the river, the City of Milwaukee Department of City Development initiated the Milwaukee RiverWalk system in 1994. Its goals were to improve public access to the downtown Milwaukee River by providing funds to establish and up-grade a network of waterfront trails, promenades, and pedestrian bridges. The system developed from a public-private partnership between property owners and the city. In exchange for permanent public access to the river, the city matches funds for private RiverWalk improvements.

The establishment of the Milwaukee RiverWalk system has attracted thousands of visitors to the downtown area and has spurred economic development along the waterside. In addition to various recreational opportunities such as RiverSplash and the Milwaukee River Challenge, property values in the RiverWalk business improvement district increased from $335 million (1994) to $517 million (2002). In addition, more than $118 million in new residential development has occurred, attracting new residents to the downtown (Milwaukee Department of City Development 2002).

Ecological Function

Much of the river corridor in the densely urban Milwaukee River South Branch watershed has been channelized, paved, or diverted underground to alleviate flooding concerns and quickly convey floodwaters downstream. These modifications have caused a marked decline in its biological diversity and ecological health. To reverse some of this damage, the Milwaukee Metropolitan Sewerage District (MMSD) is currently restoring the meandering flow of the river corridor and returning natural flood storage capacity in portions of this and other watersheds in the Milwaukee River basin.

MMSD is a state-chartered government agency providing wastewater services for twenty-eight municipalities. The district's 420-square-mile service area includes all cities and villages (with the exception of the City of South Milwaukee) within Milwaukee County and all or part of ten municipalities in surrounding Ozaukee, Washington, Waukesha, and Racine counties.

In addition to providing wastewater services, other MMSD functions include water quality research and laboratory services, operating household hazardous waste and mercury collection programs, and involvement in various environmentally focused partnerships. MMSD, in conjunction with area stakeholder groups, is charged with planning and overseeing projects to reduce the risk of flooding and protecting its sewer infrastructure and ultimately, the health of the watershed. In 1993, the EPA recognized the MMSD as a Clean Water Partner for the 21st Century in recognition of its efforts to improve the health of Milwaukee-area watersheds.

One MMSD project has focused on Lincoln Creek, a nine-mile tributary of the Milwaukee River draining a land area of approximately 21 square miles within portions of the Cities of Milwaukee and Glendale and the Village of Brown Deer (WDNR 2001a). The Lincoln Creek Environmental Restoration and Flood Control Project relocated approximately 2,025 homes and businesses out of the hundred-year (1 percent probability flood occurrence) floodplain and removed over two miles of concrete channels to restore a more natural, meandering flow (MMSD 2002).

In the 1950s, the stream was lined with concrete to convey flood surges away more quickly. MMSD has removed the concrete and restored a more natural stream. The overall habitat has improved as a result of increased natural storage capacity, and more species of fish and macroinvertebrates are gradually returning to the stream. Detention and retention basins in the watershed have increased this capacity. MMSD has done or is planning similar work along Oak Creek, Root River, and Menomonee River. Substantial changes were made along the nine-mile-long Lincoln Creek besides the removal of two miles of concrete lining, including construction of two large detention basins, improved bypass culverts and bridges, and the deepening and widening of creek segments (MMSD 2002). Although the main focus of the project is to reduce the risk of flooding, it also aims to enhance the attractiveness of the corridor; improve water quality; restore, stabilize and protect eroding banks; and provide a suitable habitat for fish, birds and other wildlife. The result is a waterway that is being viewed as a successful model for urban flood management and habitat restoration.

In addition, MMSD has implemented a land conservation plan to preserve natural ponding and undeveloped floodplain areas to help reduce the risk of future flooding. Through the assistance of a conservation fund, MMSD is working to acquire or secure easements on properties identified as critical to protecting against flooding in local watersheds (MMSD 2004).

Watershed managers are now seeking to return waterways to more natural flow regimes and allow floodwaters to disperse high-energy flows across the floodplain. This approach is particularly gaining acceptance in the downstream, urban portions of the Milwaukee River basin, where the cumulative effects of decades of structural adjustment projects such as dams, large-scale water diversions, and habitat alteration have degraded water quality. In addition, preserving or creating wetlands and protecting riparian vegetation allow for sediments and toxins to be captured and filtered before entering surface or groundwater systems. To this end, municipalities within the Milwaukee River basin are being encouraged to adopt the Regional Natural Areas and Critical Species Habitat Protection and Management Plan for Southeastern Wisconsin (SEWRPC 1997).

Future Concerns

Despite the many innovative partnerships and successful collaborative experiences in the Milwaukee River basin, there remains a need for a more comprehensive water policy management framework at a regional level to address issues confronting the entire Great Lakes region. Several agreements already exist among the eight states and two Canadian provinces adjoining the Great Lakes. One is the Great Lakes Charter, which requires the permission of all other states and provinces before allowing water withdrawals over a specified volume from the Great Lakes; another is the 1985 Toxic Substances Control Agreement between the eight states agreeing on common environmental standards to avoid unfair economic competition between them based on lax environmental regulations. In addition, the National Oceanic and Atmospheric Administration, EPA, International Joint Commission, U.S. Fish and Wildlife Service, Great Lakes Commission, and Great Lakes Fisheries Commission all have jurisdictional roles in the Great Lakes (Wisconsin Academy of Sciences, Arts and Letters 2002).

Likewise, watershed management objectives in the Milwaukee River basin must consider the entire region because the fate of the watershed's headwaters, shared aquifers, and downstream areas are inextricably linked. In the Milwaukee River basin, oversight is divided among federal, state, and local government agencies with overlapping layers of authority. The development of public-private partnerships may be of use in these circumstances. Although participation in the Milwaukee River Basin Land and Water Partners Team is currently voluntary, the group has been able to bring diverse interests together to address common concerns and improve conditions throughout the basin.

Problems still exist when considering metropolitan Milwaukee as a region, however, such as when considering regional water supply issues. While residents of Milwaukee and its older suburbs enjoy access to the abundant, fresh water from Lake Michigan, residents of burgeoning western suburbs in Waukesha County lay outside both the Milwaukee River and Great Lakes basins. Hence, they are prohibited from withdrawing water from Lake Michigan for their drinking water supply. Instead, municipalities in this area have been pumping groundwater for their drinking supply from both the shallow aquifer (approximately twenty-five to three hundred feet below ground) and from a deep sandstone aquifer. Over the past century, reliance on groundwater for household and industrial use has drawn down the latter more than six hundred feet. Of even more immediate alarm, however, is that water from this deep aquifer is enriched with naturally occurring radioactive radium, which has been linked to bone cancer, thereby threatening the health of residents in Waukesha County (Feinstein et al. 2004).

This situation is not unique to Milwaukee area residents alone. Similar circumstances exist in suburban neighborhoods within the Chicago metropolitan area. Although residents of Chicago and nearby municipalities enjoy water rights to Lake Michigan water as the result of a Supreme Court ruling, their withdrawal is limited to 2.1 million gallons of water per day. Currently, their daily intake averages 2.0 million gallons per day and frequently exceeds this allowance. In their case too, surrounding suburbs are depleting groundwater aquifers and confronting issues of high radium and other mineral concentrations.

In short, a stronger commitment to long-term regional planning and addressing both ecological and social issues within a watershed context is needed to sustain the relative health and vitality of the entire Milwaukee River basin. Together, however, the initiatives and partnerships described in this essay are indicative of steps toward this end.

Although the Milwaukee experience has followed the same historical course of watershed degradation as other U.S. cities that developed during the Industrial Revolution, new directions in watershed management, planning, and implementation focused on watershed integration are all steps in the right direction. Notably, in the case of Milwaukee's watersheds, government agencies, nonprofit organizations, and private landowners are mutually developing and implementing watershed-scale goals, management plans, and restoration projects. This involvement is significant because, although the benefits of watershed-scale restoration and management strategies are increasingly recognized throughout the United States, they are still not commonplace in practice (Dombeck, Wood, and Williams 2003).

References

Born, S. M., and K. D. Genskow. 2001. Toward understanding new watershed initiatives: A report from the Madison Watershed Workshop, 20–21 July 2000. Madison: University of Wisconsin–Madison.

Dombeck, M. P., C. A. Wood, and J. E. Williams. 2003. *From conquest to conservation: Our public lands legacy.* Washington, DC: Island Press.

Feinstein, D. T., T. T. Eaton, D. J. Hart, J. T. Krohelski, and K. R. Bradbury. 2004. *Simulation of regional groundwater flow in southeastern Wisconsin.* Wisconsin Geological and Natural History Survey Bulletin.

Franklin, J. F. 1992. Scientific basis for new perspectives in forests and streams. In *Watershed management: Balancing sustainability and environmental Change,* ed. R. J. Naiman. New York: Springer-Verlag.

Milwaukee Department of City Development, 2002. *Mayor, Historic Third Ward Association break ground for new riverwalk extension in Third Ward.* Available at http://www.mkedcd.org/news/2002/HistThirdWardRivWalk.html. Accessed on: May 16, 2002.

Milwaukee Metropolitan Sewerage District [MMSD]. 2002. Available online at http://www.mmsd.com.

NRC [National Research Council]. 1992. *Restoration of aquatic ecosystems: Science, technology, and public policy.* Washington, DC: National Academy Press.

———. 1999. *New strategies for America's watersheds.* Washington, DC: National Academy Press.

Riley, A. L. 1998. *Restoring streams in cities: A guide for planners, policymakers, and citizens.* Washington, DC: Island Press.

Rome, A. 2001. *The bulldozer in the countryside: Suburban sprawl and the rise of American environmentalism.* New York: Cambridge University Press.

SEWRPC [Southeastern Wisconsin Regional Planning Commission]. 1997. A regional natural areas and critical species habitat protection and management plan for southeastern Wisconsin. Planning Report 42.

U.S. EPA [United States Environmental Protection Agency]. 2004. Overview of current total maximum

daily load (TMDL) program and regulations. Available online at http://www.epa.gov/owow/tmdl/over
viewfs.html.

WDNR [Wisconsin Department of Natural Resources]. 2001a. The Milwaukee River Basin Fact-
sheet. PUBL# WT-719-2001. Available online at http://www.dnr.state.wi.us/org/gmu/milw/milwflyer
_801.pdf.

————. 2001b. The state of the Milwaukee River basin. PUBL# WT-704–2001. Available online at
http://www.dnr.state.wi.us/org/gmu/milw/index.htm.

Wisconsin Academy of Sciences, Arts and Letters. 2002. Waters of Wisconsin Committee Meeting
Summary—Milwaukee. Available online at http://www.wisconsinacademy.org/wow/meetings/031402
summary.html.

Watershed management efforts cannot succeed without public support of new watershed initiatives. Thus, there is a need for the public to be educated and understand the complex issues facing aquatic ecosystems. Therefore, the role of environmental education in the Milwaukee River basin is critical.

Milwaukee's commitment toward sustainability through an informed public is clear in the outreach of two remarkable environmental education and outreach programs within the city, the Urban Ecology Center and Growing Power, Inc.

The **Urban Ecology Center** is a leader in environmental education efforts in southeastern Wisconsin and is a model for other centers throughout the United States. Essentially once an abandoned park, the center was created as a part of a community revitalization effort in 1991. Situated on twelve acres of woods and riparian habitat on the east bank of the Milwaukee River and located between the most populated and diverse Riverwest and East Side communities in Milwaukee, the Urban Ecology Center is a neighborhood-based, nonprofit community center located in Milwaukee's Riverside Park.

As an outdoor laboratory, the center provides environmental education programs to local schools, promotes community environmental awareness, preserves and enhances the natural resources of Riverside Park, and protects the adjacent Milwaukee River. Each year, more than ten thousand students and teachers from twelve neighborhood schools within a two-mile radius of the center participate in the Neighborhood Environmental Education Program. Students explore their local ecology through hands-on learning experiences developed by the center's staff to complement and enrich the existing K through 12 science curriculum. In addition, the Urban Ecology Center has developed a Citizen Science Program in coordination with partners from nearby universities to conduct research within an urban environment. These programs strive to turn Riverside Park into a vibrant field station and educational facility.

Through a vigorous fund-raising campaign, the Urban Ecology Center, under the direction of executive director Ken Leinbach, has recently constructed a $5 million state-of-the-art community center (figure 2). This facility replaced a trailer that had been the center's home for over a decade. The new facility incorporates various green building technologies, such as photovoltaic and rainwater catchment systems and a green roof. (For more information, please visit the Urban Ecology Center's website at http://www.urbanecologycenter.org.)

Another Milwaukee-based organization, **Growing Power, Inc.**, is working both locally and nationally to promote increased sustainability urban agriculture (figure 3). Growing Power is a national nonprofit organization and land trust supporting people from diverse backgrounds and the environments they live in through the development of community food systems. The program provides high-quality, safe, healthy, affordable food community residents. Growing Power develops community food centers, as key components of community food systems, and offers training, active demonstration, outreach, and technical assistance. Community food centers are local places where people learn sustainable practices to grow, process, market, and distribute food.

The Growing Power Community Food Center in Milwaukee Center is the oldest working farm and greenhouse in the city. This two-acre urban farm has been continuously farmed for nearly a century. Through disseminating technical training to thousands of visitors each year, Growing Power hopes to establish local community food centers in other neighborhoods around the United States. (For more information about the program, please visit Growing Power's website at http://www.growingpower.org.)

Figure 2 The new Urban Ecology Center, Milwaukee, Wisconsin. ([*left*] photo courtesy of Sean Berry; [*below*] photo courtesy of Mark J. Heffron.)

Figure 3 Students from Chicago learn about urban agriculture from the director of Growing Power, Will Allen. (Photo by Laurin N. Sievert.)

Green Futures for Industrial Brownfields

Christopher A. De Sousa

Once viewed as symbols of urban economic power, older industrial brownfield districts located in inner cores are now perceived as little more than prime examples of urban decay. The list of socioeconomic and environmental ills associated with these districts and their surrounding neighborhoods is an extensive one and includes such "blights" as high levels of crime, crumbling infrastructure, contaminated soils, vacant buildings, "bottom-feeding" businesses, and poverty. Indeed, the physical extent of these districts and the range of the problems they face have left governments in a quandary as to what to do about them, while most city residents appear to have simply put them out of their minds.

While planners, economists, and community and business leaders discuss what can be done to revitalize these districts, a frequent theme is the increasing role that so-called greening must play in cleaning up such districts, enhancing their attractiveness for business and growth. This essay examines efforts being undertaken in three Rust Belt cities to use greening as a primary tool in the regeneration, revitalization, and restructuring of industrial brownfield districts: the Menomonee Valley in Milwaukee, the Port Lands in Toronto, and the Lake Calumet area in Chicago. These cases indicate the value of regreening as an overall strategy for the revitalization of brownfields in urban areas generally.

Brownfields

Since the early 1990s, older cities across North America have engaged in revitalizing their inner cores, most of which have been at least partially abandoned by industries, businesses, and residents. The reuse of these abandoned core districts is hampered by so-called brownfield sites, namely abandoned or underutilized properties whose past land uses have contaminated the soil or groundwater, or are perceived to have done so. Although these sites are found in all kinds of localities, both within and outside cities, they tend to be more concentrated in inner-core areas. They come in all shapes and sizes, ranging from abandoned corner gas stations to large industrial lots where manufacturing, petroleum storage, and commercial and transportation uses may have taken place. A comprehensive survey of thirty-one cities in the United States conducted by Simons (1998) estimated that in 1994 there were approximately 75,000 brownfields covering 93,500 acres and representing about 6 percent of a city's total area on average. According to a study by the U.S.

Conference of Mayors (2000), 210 U.S. cities reported having more than 21,000 brownfield sites ranging in size from a quarter of an acre to 1,300 acres. Each industrial district examined here is comprised of numerous "mixed-size" properties that cluster within single regions.

Concern over brownfields surfaced in the late 1970s. The initial focus was on finding an appropriate technology for cleaning them up and getting those responsible for creating the contamination to pay for the cleanup. Following such incidents as Love Canal, Times Beach, and Valley of the Drums, which were given broad media exposure, the federal government passed the Comprehensive Environmental Response and Liabilities Act in 1980 (CERCLA), also known as "Superfund." CERCLA made funds available for remediation and gave governments the authority to recover cleanup and damage costs from parties responsible for creating a brownfield. Fear of assuming liability, however, deterred private investors, especially banks, from becoming involved with redevelopment of any property that was remotely suspected of being contaminated. The strategy thus ended up being counterproductive, hindering efforts to remediate and redevelop many brownfield sites (Stroup 1997).

Progress was made in the mid-1990s when governments at all levels began experimenting with a range of new approaches for encouraging remediation and redevelopment (Meyer, Williams, and Yount 1995; Bartsch 1996; Simons 1998; Council for Urban Economic Development 1999). In 1995, the U.S. Environmental Protection Agency (EPA) proposed its Brownfields Action Agenda to provide funds for pilot programs, link brownfield redevelopment with other socioeconomic issues, and refocus its efforts on high-risk sites. State governments also began implementing so-called voluntary cleanup programs to promote redevelopment by offering more flexible cleanup options; giving more leeway to the private sector to oversee its own activities; and providing technical assistance, financial incentives, and protection from liability to developers and investors. At the federal level, such efforts led cumulatively to the recent passage of the Small Business Liability Relief and Brownfields Revitalization Act in 2002, which provides liability protection for prospective investors, property owners, and innocent landowners, and authorizes increased funding for state and local programs that assess and clean up brownfields.

In Canada, the federal government has always been less engaged in brownfield redevelopment, which has fallen largely under the aegis of provincial and municipal levels of government (De Sousa 2001). The general intent of governmental agencies has been to act as regulators and advisors, holding the private sector financially responsible for cleanup and redevelopment. In Ontario, the Ministry of the Environment can legally demand the remediation of a brownfield site under Canada's Environmental Protection Act. In actual fact, however, the assessment and remediation of brownfields unfolds largely as a voluntary process regulated by its Guideline for Use at Contaminated Sites in Ontario (Ontario Ministry of the Environment1996). In contrast to most U.S. jurisdictions, only in late 2002 did Ontario pass legislation designed to make investment in brownfield redevelopment more attractive to the business sector.

Overall, efforts to redevelop brownfields have produced some successes (Council for Urban Economic Development 1999; U.S. Conference of Mayors 2000). In the United States, the focus has been primarily on redeveloping brownfields for industrial and commercial uses, with residential and retail uses following closely behind. The opposite has been the order of priorities in Canadian and European redevelopment efforts (Bibby and Shepherd 1999; Box and Shirley 1999; De Sousa 2002). More recently, greater attention has been given the greening option, even though it does not directly generate significant employment or tax benefits but rather is perceived as playing an important role in improving the quality of urban life (International Economic Development Council 2001; Kirkwood 2001; De Sousa 2003).

Urban Regreening Case Studies[1]

Widespread interest in urban revitalization has led, in turn, to a resurgence of interest in greening the city (Garven and Berens 1997; Harnik 2000). So far, research on greening has focused largely on documenting the benefits and barriers associated with it. Landscape architects, for example, have focused on the aesthetic and environmental benefits that greenspace-oriented redevelopment can bestow on urban areas, such as improving environmental quality, restoring natural habitats, enhancing recreational opportunities, and improving the appearance of urban areas (Hough, Benson, and Evenson 1997; Thompson and Sorvig 2000). In addition, research has found that urban greening improves the well-being of city residents in a variety of ways, by reducing crime, reducing stress levels, strengthening neighborhood social ties, coping with "life's demands," and the like (Kuo, Bacaicoa, and Sullivan 1998; Kweon, Sullivan, and Wiley 1998; Kaplan 2001). Similar kinds of positive findings are also emerging from research conducted by environmental economists (Lerner and Poole 1999; Bolitzer and Netusil 2000). Summarizing the main implications, Lerner and Poole (1999) contend that greening projects in the United States tend to reduce costs related to urban sprawl and infrastructure, attract investment, raise property values, invigorate local economies, boost tourism, preserve farmland, prevent flood damage, and safeguard environmental quality generally.

Identifying such benefits is essential for countering the barriers, real or perceived, that are often associated with such greening, including the high maintenance costs it entails, the safety concerns it raises, and the poor accessibility it creates (Garven and Berens 1997). It is particularly true in the case of the greening of brownfield districts, and brownfields generally, which are associated with a host of socioeconomic and environmental costs and risks. Nevertheless, greening projects on brownfield sites are on the rise throughout the United States, Canada, and Europe (Garven and Berens 1997; Harnik 2000; International Economic Development Council 2001; Harrison and Davies 2002). These projects not only provide models for implementing similar works, but also highlight the important role greening plays in fashioning a more humane metropolis.

The Menomonee Valley, Milwaukee

The Menomonee Valley is a fifteen-hundred-acre old industrial corridor close to downtown Milwaukee. Prior to European settlement, the area was a diverse marsh and wetland ecosystem that provided Native Americans with a plentiful supply of fish, waterfowl, wild rice, and other resources. Starting in the nineteenth century, European settlers were attracted to the valley by its transportation potential, given its location at the confluence of two major rivers, the Milwaukee and the Menomonee, that converge at the city center and flow into Lake Michigan. Canals, roads, and water and sewer systems were constructed which attracted industrial interests to the city (figure 1). By the 1920s, more than fifty thousand people were employed by these economic enterprises in the valley.

Industrial decline in the Menomonee Valley started during the Great Depression of the 1930s and became widespread by the late 1970s. In addition to job losses, the decline turned the valley into Wisconsin's largest brownfield site, laden with polychlorinated biphenyls, heavy metals, petroleum residue, and other contaminants typical of former industrial activities (State of Wisconsin Brownfields Study Group 2000). Although the remaining businesses in the valley still employ more than seven thousand people, its contamination problems, both real and perceived, continue to pose a daunting and complex challenge to any redevelopment scheme.

Figure 1 Milwaukee's Menomonee River Valley, 1882. (Source: Historic Urban Plans 1978.)

The City of Milwaukee and key stakeholders have joined forces to devise ways to rekindle the industrial potential of the valley and revitalize its natural resources. The Menomonee Valley Partners, a public-private partnership bringing together members of the business world, community organizations, and government agencies, was established to facilitate the implementation of the City of Milwaukee's Land Use Plan for the valley. On the whole, the Menomonee Valley Partners (2003: homepage) envision a redeveloped Valley that is as central to the city as it was in the past:

- Geographically central, with new ties to the surrounding neighborhoods;
- Economically central, with strong companies that provide jobs near workers' homes;
- Ecologically central, with healthy waterways and greenspace; and
- Culturally central, with recreational facilities for the community.

All levels of government in Wisconsin are now making available an extensive array of financial incentives to prospective developers. There is also an ongoing planning process designed to protect the valley's natural resources and restore some of its previous habitat and natural systems.

The first significant greening project initiated in 1992 was the Hank Aaron State Trail, which was officially opened in 2000 on the valley's west side. When completed, the trail will be a seven-mile urban greenway through the heart of the valley (figure 2). The primary objectives of the project are

- Protection and renewal of the riparian corridor
- Development of a multiuse pathway for commuting
- Provision of close-to-home recreational activities for adjacent neighborhoods
- Use of the valley for its historical value
- Linkage of the trail to other city, county, and state trail systems.

As mentioned, the Department of Natural Resources of Wisconsin is the lead agency in planning, implementing, and managing the project. The City of Milwaukee is involved primarily in raising funds, releasing land, and maintaining the trail itself. Various federal agencies have provided financial support for accessories such as signs and artwork. Local community groups and neighborhood associations such as the Friends of the Hank Aaron State Trail have helped raise awareness and funds while assisting with special events. Private landowners (e.g., Miller Park Stadium Corporation) are being contacted by the state to donate easements for the trail and help finance development and renaturalization activities. The Department of Natural Resources estimates that the total project costs will amount to slightly over $5 million, with open space costing approximately $450,000 and site

Figure 2   Hank Aaron State Trail, 2002. (Photo by C. De Sousa.)

assessment and cleanup $500,000. The remaining funds will go toward site acquisition, project planning, and site development.

One local nonprofit group, the Sixteenth Street Community Health Center, together with the City of Milwaukee and other sponsors, organized a national design competition, *Natural Landscapes for Living Communities,* to plan the redevelopment and greening of a 140-acre abandoned railroad property in the western end of the Menomonee Valley within the city. The aims of the competition are implicit in the criteria presented to the four finalist design teams (Sixteenth Street Community Health Center 2002):

- To design an industrial park accommodating at least 1.2 million cubic feet of development (proposed by the city)
- To extend Canal Street (a major connection road within the valley)
- To expand the Hank Aaron State Trail
- To interconnect the railroad property to Mitchell Park and neighborhoods to the north and south of the valley
- To devise site-specific storm and flood water management techniques
- To resolve site-specific environmental and geotechnical issues
- To landscape the area

• To establish community connections to the site by means of open space planning, educational opportunities, and signage.

The preliminary vision for the site put forward by Wenk Associates of Denver,
Colorado (selected in the summer of 2002) incorporates the full range of criteria
listed above. Their design includes an industrial park surrounded by a variety of
natural and open space features, including a "storm water" park, trails, a community green space, and a renaturalized Menomonee River (figure 3). In all, the 140-
acre site is slated to encompass 70 acres of light industry; a one-mile segment of
the Hank Aaron State Trail; and 70 acres of streets, parks, and natural areas along
the banks of the Menomonee River (Wenk Associates 2002). The city is currently
in the process of preparing the site for redevelopment, while other stakeholders are
raising both awareness and funds to ensure that the project continues to move
forward.

The Toronto Port Lands

The revitalization of Toronto's Port Industrial District, often referred to as the Port
Lands, has been the subject of intense debate for more than two decades. Located
southeast of the central business district, the one-thousand-acre property was created largely by fill from dredging, demolition, and other such activities in the city.
Currently, there is a range of industrial, commercial, and recreational uses on the
Port Lands, including Toronto's port facilities. Historically, the energy companies
occupied a large portion of the area, with oil tank farms making up almost a half of
the total area (Hemson Consulting 2000). The energy crises of the 1970s and the
subsequent switchover to natural gas for residential energy led to a decline in the
need for oil, which led, in turn, to the migration of oil companies away from
the port area. Although more than three thousand people still work for businesses
located in the Port Lands, the site is becoming gradually abandoned and is extensively contaminated, containing more than one hundred individual brownfield
sites (Hemson Consulting 2000; Groeneveld 2002).

The debate over the future of the Port Lands has always been a heated one. Some
interests believe that it is best suited for residential redevelopment, bringing the
district in line with other successful residential communities along the waterfront
(Warson 1998). Others envision the area as a continuation of the larger greenspace
renewal efforts that have been taking place in contiguous areas to the port (the
Don River to the north, the Leslie Street Spit and Cherrie Beach to the south). Finally, some believe that the area should continue to be used for commercial and
industrial uses. All agree, however, that some form of greening must take place as
part of any viable revitalization scheme for the area.

The first comprehensive attempt at developing a greening plan for the Port
Lands was undertaken by the Waterfront Regeneration Trust, an agency that grew

This plan shows a slight realignment of the parking ring road and Miller Park East parking as a potentially desirable future condition. Neither the realignment of the road, nor the realignment of the parking will affect the constructability of the plan, and no proposed improvements are dependent on these future actions.

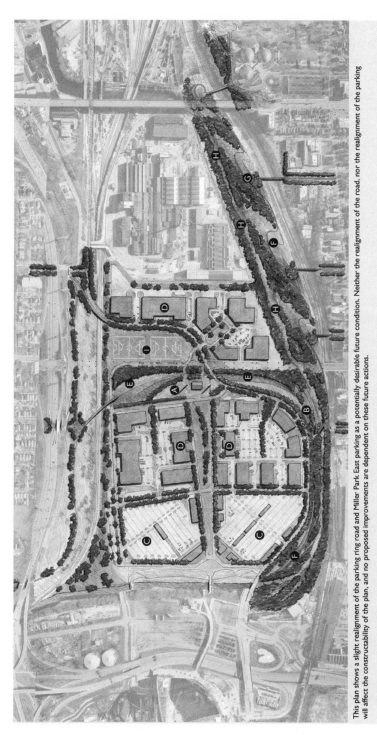

Ⓐ COMMUNITY GREEN

Ⓑ STORMWATER PARK

Ⓒ MULTI-PURPOSE OUTDOOR EVENT SPACE

Ⓓ ECO-INDUSTRIAL PARK

Ⓔ STORMWATER PARK DETAIL

Ⓕ HANK AARON STATE TRAIL AND NATURAL AREA

Ⓖ BLUFF OVERLOOK

Ⓗ RIVER POINTS

Ⓘ SOCCER FIELDS

NORTH

200 0 200 400

SCALE IN FEET

Figure 3 Landscape Concept Framework, Menomonee River Valley Design Competition, Wenk Associates, 2002. (Courtesy of Sixteenth Street Community Health Center, Milwaukee.)

out of a Royal Commission established in 1988 to study the future of the Toronto waterfront. The trust adopted an ecosystem approach that integrates community, environmental, and economic needs into the redevelopment of contaminated lands. In 1997, the trust published *Greening the Toronto Port Lands,* which contained a plan for green infrastructure for the Port Lands (Hough, Benson, and Evenson 1997).

More recently, greening of the Port Lands was used as a tactic by the City of Toronto in its bid for the 2008 Summer Olympics. Although the bid failed, a Waterfront Revitalization Corporation (WRC) was established nonetheless to move redevelopment and renewal activity forward. According to the plan developed (Toronto Waterfront Revitalization Task Force 2000), new public spaces will encompass 450 acres throughout the waterfront, and the WRC has pledged $500 million dollars specifically for park development. As for the port itself, the plan provides for an extension of its greenspace from 5 percent to 30 percent of the total area to provide more habitat, improve the ecology of the region, provide recreation, and manage storm water. A "green border" is slated to surround the port to renaturalize the waterfront and allow the public easy access to it. It is also anticipated that the port will accommodate approximately twenty-five thousand new homes and numerous "new economy"-oriented businesses (those involved in information technology, media, biomedical and biotechnology, and professional services). In total, the anticipated twenty-year renewal of the waterfront will cost an estimated $12 billion Canadian, of which over $5 billion will come from public sources to cover site acquisition, infrastructure, and business interruption/relocation costs. The WRC is responsible for raising the remaining funds via public/private partnerships. Thus far, the three levels of government have pledged $1.5 billion of initial funding for a variety of so-called priority projects, including the cleanup of contamination that is estimated to cost between $60 million to $500 million, depending on the approach taken.

One such priority project is "restoring" the mouth of the Don River, where the port and the river meet (figure 4). According to the WRC (2000, 1), "The green corridor is intended to serve as a welcoming entrance to the Port Lands and encourage private sector investment and future development." The project will connect Toronto's waterfront to greenspace in the Don River Valley, transforming vacant lots and concrete into fifty-two acres of new parkland, wetland, and marsh areas. It will also improve water quality and free up new land for redevelopment in the West Don Lands, an industrial brownfield area located just north of the port and often considered to be an extension of the Port Lands for planning purposes. Fulfilling this vision will require extensive soil and groundwater remediation, removal of current infrastructure, and the reconfiguration of the mouth of the Don River. The WRC has already set aside $2 million (Canadian) for the assessment, design, and planning process itself (Toronto Waterfront Revitalization Corpora-

Figure 4 The Mouth of the Don River. (Photo by C. De Sousa.)

tion 2000). The project is envisioned as being another successful brownfields-to-greenspace project that Toronto has undertaken within its central city and in areas surrounding the Don River since the early 1990s (see De Sousa 2003).

Chicago's Calumet District

The Calumet region on the far south side of Chicago is a classic example of a planning exercise that sees industrial and natural concerns as complementary. The area contains beaches, marshes, moraines, ponds, and slow-moving rivers (U.S. National Parks Service Midwest Region 1998). In the 1840s, railroads traversed the region and people started settling into the area. As shipping activity increased in the Great Lakes, industrialization and urbanization expanded in the Calumet area. Throughout the twentieth century, the steel industry was the main user of the land and shaper of the local culture. Inevitably, substantive quantities of wastes were deposited throughout the region. By the mid-1970s, the steel production industry in Calumet began to falter owing to a decline in steel use. The subsequent closing of mills in the area had a devastating effect on the local neighborhoods that supported them.

Alongside its industrial activities, the Calumet region has always retained, in part, a rich ecological and recreational character. Given its location in Chicago, the area has made for excellent hunting, fishing, and recreation for Chicagoans. Despite its industrial history, the region still possesses numerous natural areas: extensive prairie districts, dunes, and wetlands that provide a rich habitat for plants and wildlife, including many rare and endangered species. Calumet is also famous among birdwatchers because of the thousands of bird species that fly to the region during the spring and fall migrations (City of Chicago 2002; Darlow 2002). As the U.S. National Parks Service Midwest Region (1998) aptly put it, "Today, the Calumet region exists as a unique mosaic of globally rare natural communities and significant historic features in juxtaposition with heavy industry."

Renewal of the area has been the target of extensive debate and study since the 1970s (City of Chicago 2002). As in Milwaukee, a grant from the EPA helped initial efforts research and plan a sustainable future for the area. Of these plans, the *Calumet Area Land Use Plan and Calumet Open Space Reserve Plan* (December 2001) proposed by the City of Chicago received the most attention. This plan focuses on a five-thousand-acre section of the Lake Calumet bioregion that covers the city's south side. (In 1996, the National Parks Service initiated the Calumet Ecological Planning Study to examine the entire Calumet region for potential addition to the National Parks System. The study area encompasses portions of Porter and Lake counties in Indiana in addition to the area within Illinois.)

The objectives of the plan are as follows:

- To improve the quality of life in the Calumet area and the surrounding communities by creating greater economic opportunities and enhancing environmental quality
- To retain and enhance existing businesses and industries within the Calumet area
- To attract new industrial and business interests
- To create new job opportunities
- To protect and revitalize wetland and natural areas within the Calumet area and improve habitat for rare and endangered species.

Of the five-thousand-acre planning area, one thousand acres of largely former manufacturing brownfield sites have been set aside for industrial redevelopment. Such redevelopment will be supported through financial incentives from tax increment financing and from projects designed to upgrade the transportation infrastructure of the area. The remaining four thousand acres will be used largely as greenspace, habitat, and so-called reclaimed space (greenspace on land that was used formerly for waste disposal).

One of the plan's initial projects foresees linking greenspace with industrial and neighborhood renewal on the former South Works Steel Mill site. In its heyday, more than twenty thousand people worked at the 573-acre lakefront site located at the mouth of the Calumet River. The mill began shutting down operations in phases in the 1970s and closed completely in 1992. The owner, USX Corporation of Pittsburgh, voluntarily completed cleanup at the site in 1997 to meet residential standards. Planning started in 1999, with the main partners being the City of Chicago, the Chicago Park District, the Department of Transportation, USX, and various private developers. The plan envisions a lakefront park that will connect it to the system of open spaces, parks, and civic spaces along the Chicago waterfront. Extensive habitat improvements will be made to the mouth of the Calumet River, and active recreation facilities will be constructed in the northern portion of the site. At the same time, residential development and an industrial waterfront will be created on which modern manufacturing sites, warehouses, and offices will be constructed. Although costs are still being determined, the plan is anticipated to follow in the footsteps of other successful efforts to renew Chicago's waterfront.

Implications

The industrial brownfield districts described here are examples of emerging planning success stories in a postindustrial world. The brownfields of these districts are unique in that they present similar barriers and opportunities to planning for a humane metropolis. Once sought after for their resources and transportation linkages, the legacy of heavy industrial use on these lands has left deep scars in the landscape. The very contaminated soil and groundwater spoil that characterize these districts are extremely costly to remediate compared with other kinds of brownfield lands. In most of the cases, the costs must come primarily from the public purse because prior landowners either no longer exist or are bankrupt. The outdated buildings and infrastructure that have kept new businesses away for decades require costly removal or significant upgrading. Politically, efforts to plan a viable future for these sites have often been mired in jurisdictional clashes and in contrasting viewpoints on the part of numerous interested parties. And, on the environmentalist side, these districts have often been perceived to be barren wastelands that are beyond recovery, making it difficult to get funding for greening purposes.

The three case studies examined here, however, show that such barriers can be overcome. These examples constitute opportunities for turning wastelands into success stories. Above all else, they present contexts for partnership alliances that can be forged among the many disparate interest groups that make up the sociopolitical arena. Businesspeople, governmental agencies, community groups, landowners, and environmentalists are now starting to understand that renewal of such prime districts can only come about through a sharing of the burdens of redevelopment. In addition, from such brownfield redevelopment successes as those reported here there is a growing feeling among planners that the partnership model has broader applicability. The districts described are now becoming exemplars for redevelopment of brownfield districts on a larger scale.

Greening in particular is being perceived more and more as a way to restore such sites to what they were before industry polluted them. Unlike projects that aim to develop small brownfield sites on their own, as autonomous redevelopment schemes, the case studies reported here show that it is much more preferable to integrate such sites into a framework for redevelopment of the entire district that encompasses them. In this way, a multitude of economic, social, and environmental renewal objectives can be achieved simultaneously.

It has become clear that greening and brownfield redevelopment are two sides of the same coin in any effort to humanize the metropolis. Nowhere has it become more apparent than in the revitalization of industrial brownfield districts such as those in Milwaukee, Toronto, and Chicago. Along with comparable redevelopment projects in North America and Europe, they are particularly useful as models for helping cities develop appropriate renewal schemes for their previously designated industrial sites. In a postindustrial society, the individualistic approach to renewal is, simply put, not the way to go. Partnership among previously conflicting groups is the path to building the humane metropolis.

Note

1. Information for this section was obtained from a review of planning documents published by the cities of Milwaukee, Toronto, and Chicago and from site visits. The districts examined have been the target of extensive planning and some preliminary redevelopment and greening activity. For each district, information on a specific redevelopment/greening project was obtained through a survey questionnaire. Rather than provide an in-depth data analysis, the purposes here are to assess the potential effects of the three case study districts and derive implications from them in a more general way.

References

Bartsch, C. 1996. Paying for our industrial past. *Commentary* (Winter): 14–24.

Bibby, P., and J. Shepherd. 1999. Refocusing national brownfield housing targets. *Town and Country Planning* 68(10): 302–6.

Bolitzer, B., and N. R. Netusil. 2000. The impact of open spaces on property values in Portland, Oregon. *Environmental Management* 59:185–93.

Box, J., and P. Shirley. 1999. Biodiversity, brownfield sites and housing. *Town and Country Planning* 68(10): 306–9.

City of Chicago. 2002. *Calumet open space reserve plan.* Chicago: City of Chicago, Department of Planning and Development.

Council for Urban Economic Development. 1999. *Brownfields redevelopment: Performance evaluation.* Washington, DC: Council for Urban Economic Development.

Darlow, G. 2002. Lake Calumet: Where industry and nature meet. *The Field* 73(4): 4–6.

De Sousa, C. 2001. Contaminated sites management: The Canadian situation in an international context. *Journal of Environmental Management* 62(2): 131–54.

———. 2002. Brownfield redevelopment in Toronto: an examination of past trends and future prospects. *Land Use Policy* 19(4): 297–309.

———. 2003. Turning brownfields into green space in the City of Toronto. *Landscape and Urban Planning* 62(4): 181–98.

Garvin, A., and G. Berens. 1997. *Urban parks and open space.* Washington, DC: Urban Land Institute and Trust for Public Land.

Groeneveld, T. 2002. *Navigating the waters: Coordination of waterfront brownfields redevelopment.* Washington, DC: International City/County Management Association.

Harnik, P. 2000. *Inside city parks.* Washington, DC: Urban Land Institute and Trust for Public Land.

Harrison, C., and G. Davies. 2002. Conserving biodiversity that matters: Practitioners' perspectives on brownfield development and urban nature conservation in London. *Journal of Environmental Management* 65:95–108.

Hemson Consulting. 2000. *The port area of Toronto: Maintaining a valuable asset.* Toronto: Report prepared for Lafarge Canada Inc.

Historic Urban Plans. 1978. Milwaukee Wisconsin in 1882. Ithaca, NY: Copied from a lithograph produced in 1882 by Beck & Pauli in the Library of Congress.

Hough, M., B. Benson, and J. Evenson. 1997. *Greening the Toronto Port Lands.* Toronto: Waterfront Regeneration Trust.

International Economic Development Council. 2001. *Converting brownfields to greenspace.* Washington, DC: International Economic Development Council.

Kaplan, R. 2001. The nature of the view from home. *Environment and Behavior* 33(4): 507–42.

Kirkwood, N., ed. 2001. *Manufactured sites.* London: E & F Spon.

Kuo, F. E., M. Bacaicoa, and W. C. Sullivan. 1998. Transforming inner-city landscapes: Trees, sense of safety, and preference. *Environment and Behavior* 30(1): 28–59.

Kweon, B., W. Sullivan, and A. Wiley. 1998. Green common spaces and the social integration of inner-city older adults. *Environment and Behavior* 30(6): 832–58.

Lerner, S., and W. Poole. 1999. *The economic benefits of parks and open space.* Washington, DC: Trust for Public Land.

Menomonee Valley Partners. 2003. *Menomonee Valley Partners Home Page.* Available online at http://www.renewthevalley.org/.

Meyer, P., R. Williams, and K. Yount. 1995. *Contaminated land: Reclamation, redevelopment, and reuse in the United States and European Union.* Aldershot, UK: Edward Elgar.

Ontario Ministry of the Environment. 1996. *Guideline for use at contaminated sites in Ontario.* Toronto: Queen's Printer for Ontario.

Simons, R. 1998. *Turning brownfields into greenbacks.* Washington, DC: Urban Land Institute.

Sixteenth Street Community Health Center. 2002. Menomonee River Valley national design competition, executive summary. Milwaukee: Competition sponsored by the Sixteenth Street Community Health Center, Menomonee Valley Partners Inc., the City of Milwaukee, the Milwaukee Metropolitan Sewerage District, Wisconsin Department of Natural Resources, and Milwaukee County.

State of Wisconsin Brownfields Study Group. 2000. *Brownfields Study Group final report.* Madison: Wisconsin Department of Natural Resources.

Stroup, R. L. 1997. Superfund: The shortcut that failed. In *Breaking the environmental policy gridlock,* ed. T. L. Anderson. Stanford, CT: Hoover Institution Press.

Thompson, J. W., and K. Sorvig. 2000. *Sustainable landscape construction: A guide to green building outdoors.* Washington, DC: Island Press.

Toronto Waterfront Revitalization Task Force. 2000. *Our Toronto waterfront.* Toronto: Toronto Waterfront Revitalization Task Force. Available online at http://www.city.toronto.on.ca/waterfront/fung_report.htm.

U.S. Conference of Mayors. 2000. *Recycling America's land: A national report on brownfields redevelopment,. Vol. 3.* Washington, DC: United States Conference of Mayors.

U.S. National Parks Service Midwest Region. 1998. *Calumet Ecological Park feasibility study.* Washington, DC: National Park Service. Available online at http://www.lincolnnet.net/environment/feasibility/calumet3.html.

Warson, A. 1998. Toronto's waterfront revival. *Urban Land* (January): 54–59.

Wenk Associates. 2002. A vision for the Menomonee River Valley. Prepared for the Menomonee River Valley national design competition, Milwaukee, Wisconsin, sponsored by the City of Milwaukee Department of Environmental Health, the Sixteenth Street Community Health Center, and other stakeholders.

WRC [Waterfront Revitalization Corporation]. 2002. Priority projects. Available online at http://www.towaterfront.ca.

Ecological Citizenship

The Democratic Promise of Restoration

Andrew Light

The writings of William H. Whyte do not loom large in the literature of my field: environmental ethics, the branch of ethics devoted to consideration of whether and how there are moral reasons for protecting nonhuman animals and the larger natural environment. Environmental ethics is a very new field of inquiry, only found in academic philosophy departments since the early 1970s. Although there is no accepted reading list of indispensable literature in environmental ethics, certainly any attempt to create such a list would begin with Henry David Thoreau, John Muir, Aldo Leopold, Rachel Carson, and a more recent handful of senior scholars who had been writing on these topics early on, such as J. Baird Callicott, Val Plumwood, Peter Singer, Richard Sylvan, Tom Regan, and Holmes Rolston III (for a review of contemporary environmental ethics, see Wenz 2001; Light 2002; and Palmer 2003).

Environmental ethics aims to be an interdisciplinary endeavor. As such, the required reading list in this field should be more open than the traditional philosophical canon, inclusive of those environmental thinkers who either were not philosophers or whose philosophical status is a matter of some dispute. Such a claim is evidenced by the short list just recited: included there are figures like Leopold who, while trained as a professional forester, arguably wrote one of the most important foundational works for environmental ethicists, the penultimate chapter of his autobiographical *A Sand County Almanac,* "The Land Ethic." In thinking about the recent history of the development of this field of inquiry, however, the gaps in who is considered to be indispensable for those new to the field seem more important than who would be included.

Much of my own work in environmental ethics has been devoted to the claim that the field is failing as a discipline that has much to say about the actual resolution of environmental problems. A considerable amount of literature on environmental ethics is focused on questions of the abstract value of nature as it is found

This essay is a shortened and revised version of my "Restoring Ecological Citizenship" in *Democracy and the Claims of Nature,* ed. B. Minteer and B. P Taylor (Lanham, MD: Rowman and Littlefield, 2002), pp. 153–72. Consult the original version of this essay for the full prosecution of the argument presented here.

in its most pristine form, namely wilderness. Most of the contemporary philosophers listed above (excluding animal welfare advocates like Singer and Regan) have primarily focused their work on providing arguments for wilderness preservation, or at least on questions of natural resource conservation found outside of densely populated areas (see Light 2001). Rarely, if ever, do environmental ethicists discuss how to form better relationships between society and nature in human-dominated settings—namely cities or other urban communities—rather than simply considering the value of nature in the abstract. Surely the blindness to urban issues is arguably in part a reflection of the larger antiurban tendencies of the broader environmental community.

Thus, it is not surprising that the writings and ideas of William H. Whyte are conspicuously missing from the standard reading list of environmental ethics. If they were, then their inclusion would suggest that environmental ethicists pay attention to an entirely different set of questions than those that most of the senior scholars in the field are concerned with. The same applies to Lewis Mumford, Jane Jacobs, and other nonphilosophers who raise important ethical questions about the human habitat and the design of urban space.

I am convinced that Whyte should be on the reading list of every environmental philosopher, regardless of the focus of his or her work. There are many reasons, but perhaps most important is that Whyte was concerned more with the "nature," or, rather, the open spaces, that most of us will encounter in our daily lives—the strips of land here and there near our homes—than with the great wilderness areas that most people will never see. He did not have this focus out of mere predilection but because he knew that these smaller bits of land—"tremendous trifles" as he put it—were in the end more important to the everyday lives of people than the spaces farther afield. If Whyte is correct, and if environmental ethics as a discipline is concerned with our possible moral responsibilities to the land around us, then paying attention to Whyte's work could help redirect the geographical focus of environmental ethicists to a field of inquiry more relevant to the interests of most people.

Although Whyte was not an ecologist, his reasons for this focus are entirely consistent with a sound human ecology of how people should live in relation to the broader natural environment. Whyte was a preeminent champion of the importance of density as the only sane future for land use policy in America. He worked hard to try to show how density was better for us, and the land around us, and how it could be improved to make it more attractive as an alternative to the growing sprawl that he documented so well and countered in *The Last Landscape* (Whyte 1968). None of that was to argue that wilderness preservation, conservation of species biodiversity, or the like were not important environmental priorities, but rather to raise awareness that just as important is our relationship to one another as it is mediated by the nature closer to home.

Such concerns led Whyte to focus as much on the perception of open space as the physicality of it or, as he put it, two kinds of reality: "One is the physical open space; the other is open space as it is used and perceived by people. Of the two, the latter is the more important—it is, after all, the payoff of open-space actions" (Whyte 1968, 165). For Whyte, the brook by the side of the road was just as important, if not more important, than the grand plans for regional parks. This focus speaks to a fundamental insight by Whyte that most philosophers working in environmental ethics have forgotten or indeed never paid heed to at all: *that our relationship to nature is ultimately shaped locally.* It is therefore in our immediate backyards—streets, parks, stream banks, and remnants of woods, prairie, or desert—that we must demonstrate the importance of natural amenities to people if we ever hope to show them the importance of larger environmental questions. Eventually, there should be compatibility between the two; the local environment that comes to be cared for and loved by its neighbors becomes a reason for concern with larger scales of ecological phenomena. In our quest to articulate the value of nature itself, absent its modification by humans, however, philosophers at least have forgotten that the natural spaces that we do in fact inhabit make up the "last landscape" of most immediate importance.

Such intuitions have driven my work toward those environmental practices that tend to encourage a kind of stewardship or, more precisely, "ecological citizenship" between people and the land around them. Much of this work has focused on restoration ecology as one practice that can help reconnect people to the land. Regrettably, other environmental ethicists have decried restoration as "faking nature" (Elliot 1997) that either has no place in an ethical form of conservation or at best is secondary to larger schemes of preservation. Yet in restoration I have seen what Whyte saw in the tremendous trifles that he called our attention to so well.

In this light, I will first offer a brief explanation of what restoration ecology is, its importance, and the ethical dimensions of its practice. Next, the arguments for public participation in restoration will be reviewed. Then, one possible model for framing this participation—ecological citizenship—will be proposed. Finally, some relevant public policy implications will be identified. Although the original formulation of these ideas did not rely on a reading of Holly Whyte, I now see it as a consistent extension of important themes in his work. I do not think that this influence is accidental, but rather proof of the continuing influence of Whyte on the community of scholars, activists, and policy makers who have shaped the environmental context out of which this work has been produced.

Ethics and Restoration Ecology

Restoration ecology is the practice of restoring damaged ecosystems, mostly those that have been disturbed by humans. Such projects can range from small-scale urban park reclamations, such as the ongoing restorations in Central Park and Prospect Park in New York City, to huge wetland mitigation projects as in the Florida Everglades. Restoration ecology is becoming a major environmental priority, in terms of number of voluntary person-hours devoted to it and amount of dollars committed to it by public and private sponsors. For example, the cluster of restorations coordinated by the regional network, "Chicago Wilderness," in the forest preserves surrounding Chicago (discussed more below), attracted thousands of volunteers to help restore more than seventeen thousand acres of native oak savannah (Stevens 1995; Gobster and Hull 2000). The final plan for the Chicago Wilderness program is to restore upwards of one hundred thousand acres. In the same region, the City of Chicago is committing an estimated $30 million to restoring selected wetlands within the industrial brownfield region at Lake Calumet on the city's south side (see the essay by Christopher A. De Sousa, this volume).

In Florida, various government agencies have spent hundreds of millions of dollars on returning the Kissimmee River to its earlier meandering path (Toth 1993). Work on the Kissimmee and other watersheds in Florida has revealed that even more extensive restoration is needed to fully address the threats caused by channelization to water reserves, endangered species, and the Everglades ecosystem. A plan submitted by the Clinton administration and approved by Congress in 1999 appropriated $7.8 billion of funding over the next twenty years to restoring the Everglades, making it one of the largest pieces of environmental legislation in U.S. history (Wald 1999).

Ecological restorations can be produced in a variety of ways. Although the Chicago restorations have involved a high degree of public participation, others have not. Partly, the differences in these various projects have been a result of their differing scale and complexity. Dechannelizing the Kissimmee River is a task for the Army Corps of Engineers (which, after all, channelized it in the first place) and not a local community group. Many restorations that could conceivably involve community participation, however, often enough do not, and some that already involve community participation do not use that participation as much as they could.

The alternative to community participation is to hire a private firm or use a government service to complete the restoration. One need only scan the back pages of a journal such as *Ecological Restoration* (formerly *Restoration and Management Notes*, one of the main journals in the field) to see the many landscape design firms and other businesses offering restoration services.

One important question is, Which method should be used to conduct a restoration project where options are available: volunteers or professional contractors? The answer depends in part on what we hope to achieve in any particular restoration. Most restorations are justified in terms of increasing the ecosystemic health of a landscape or restoring a particular ecosystem service or function. In such a case, most people will argue that the ends should justify the means: we should use the most efficient scientific means to achieve a desired end, namely a professional firm or a government agency specializing in such work.

Such an approach, though, assumes that the only relevant criteria for what counts as a good restoration are scientific, technological, design, and economic factors. *There is also an important moral dimension to a good restoration, namely the degree of public participation involved in such projects* (Light and Higgs 1996; Light 2000c). This view argues that there are unique values at stake in any restoration that can be achieved only through some degree of public participation in a project, for example, the potential of restorations to help nurture a sense of stewardship or care between humans and the nature around them. Such social or moral values to the community augment the other values of restoring the ecological condition of a site per se.

To achieve these moral values, a good restoration should maximize the degree of hands-on public participation appropriate for a project, taking into consideration its scale and complexity. Ideally, volunteers should be engaged in all aspects of a project, including planning, clearing, planting, and maintenance. This public participation does not mean that expertise should be abandoned in restorations; it just means that whenever possible, restorations are better when experts help guide voluntary restorationists. Based on such arguments, I have claimed that the practice of restoration ecology is as much about restoring the human relationship with nature as it is about restoring natural processes themselves. Not to attempt to achieve both of these ends in restorations is to lose the potential moral benefits of restoration.

What kind of participation is best for a restoration? I suggest that a democratic model of participation, which I call "ecological citizenship," is the best model for achieving the full potential of restoration in moral and political terms. How we shape practices and policies involving restoration is a critical test for how deep a commitment to encouraging democratic values we have in publicly accessible environmental practices. Before explaining this point, though, let us consider the simpler participatory benefits of restoration.

Restoration and Democratic Participation

Several arguments have been put forward for the importance of democratic participation in environmental decision making. According to Sagoff (1988), access to environmental amenities should be understood in the United States at least as a right of citizenship rather than only as a good to be consumed. Public participation in the formation of environmental policy was given perhaps its strongest empirical defense in Adolf Gundersen's study (1995) demonstrating the positive environmental consequences of democratic decision making. Contrary to many expectations, Gundersen argued that opening environmental decisions to the public does not necessarily weaken those decisions and in many ways may make them stronger. More recently, other philosophers and political theorists have made specific proposals for democratic environmental reforms, such as environmental constitutional rights, environmental trusteeship, and methods for expanding environmental justice (see the essays in Light and de-Shalit 2003).

All these scholars—let us call them "democratic environmental theorists"—rely on a set of common premises. The first is that environmental ethicists and political theorists must accept the democratic context of environmental decision making in which we in the developed world (and largely in international institutions) find ourselves. There is no room among these scholars to consider Malthusian arguments that would force some form of "green totalitarianism" on people. Second, these theorists all assume that it would be better to go further and actively endorse and expand the democratic context of environmental decision making because in the end it will provide the basis not only for better forms of environmental protection but also better human communities as well, helping bring people together in stronger social networks.

Following from the first premise, it is proposed that only a democratic environmentalism can actually achieve long-term sustainability. Such a position conflicts with most approaches in environmental ethics by considering the traditional ways that humans value nature (e.g., aesthetic value, resource value, or the value of protecting the environment for future generations) in contrast with the view that nature only has moral status if it has some form of noninstrumental or intrinsic value. Something is said to have intrinsic value when it is valuable in and of itself without reference to its value for other ends. To attribute such intrinsic value to nature resembles classical ethical arguments for why humans are the kinds of beings to which we owe moral obligations. For example, Immanuel Kant (1785) famously argued that humans possess special properties such that we should never reduce them solely to the value they have to us to help achieve our own ends. We should try to respect all persons as an end unto themselves and so should grant them at least some minimal level of moral respect.

Most environmental ethicists postulate a similar value for nature, namely to esteem nonhuman species and ecosystems regardless of their instrumental or economic value solely to humans. Such a view resists appeals only to human interests as a basis for valuing some bit of nature, in part because such arguments cannot guarantee that nature will be protected against competing claims for a human interest in exploiting or developing nature.

One problem, however, is that such views may degenerate into the complacent assumption that compliance with a moral principle will follow if the principle can be shown to be theoretically justified. If traditional environmental ethicists can provide the rationale for the intrinsic value of nature, then it is assumed that people will eventually act accordingly and come to respect nature in a moral sense. Yet there are precious few good reasons to accept such a view. Just because a moral reason can be offered, and even defended as true, does not guarantee that it will be followed. The more important question is, What sorts of reasons would morally motivate someone to change his or her behavior for the betterment of nature? This question requires going beyond abstract discussions of the value of nature to consider instead, for example, which practices might encourage an embrace of the importance of the long-term sustainability of the environment. Another way of putting the same point would be to ask, What practices make people better stewards of the environment?

Encouraging a direct participatory relationship between local human communities and the "nature" around them is one way to elicit such a sense of stewardship. Communities that have a participatory relationship with the land around them are less likely to allow it to be harmed, in contrast with "top-down" regulations or mandates from a higher authority that may be ignored or opposed locally (see Curtin 1999 for some examples). Noting that three-quarters of the American people live in metropolitan areas, urban ecologist Steward T. A. Pickett (2003, 67) puts it this way: "If the public bases its understanding of ecological processes on its local environment, then extracting ecological knowledge from urban systems has the best chance of enhancing ecological understanding worldwide."

Restorations performed by volunteers arguably tend to foster these kinds of relationships. For instance, a study of 306 volunteers in the Chicago restoration projects reported that the respondents were most satisfied with a sense of meaningful action ("making life better for coming generations" or "feeling that they were doing the right thing") and fascination with nature ("learn how nature works") (Miles, Sullivan, and Kuo 2000, 222). Listed third behind those values was participation (e.g., helping people feel they were "part of a community" or "accomplishing something in a group"). This study also found that length of experience in restoration activities was not a significant factor in whether people gained such perspectives: Although the length of involvement of the 306 respondents ranged from two months to twenty-seven years, "the benefits an individual derived from restoration were the same whether the individual was a relatively recent recruit or an 'old hand' " (Miles, Sullivan, and Kuo 2000, 223).

This study and others (see, e.g., those in Gobster and Hull 2000) indicate that participation in restorations has the potential for promoting strengthened attitudes toward long-term sustainability through appeal to human interests and thus may produce better connections between people and nature in places closer to home. In the context of the views of the democratic environmental theorists, however, there is more work that could be done here. Restorations clearly have the potential for producing good environmental stewards who feel a close personal connection to the land that they have come to care for. But what about a more ambitious notion of participation than that implied by "stewardship"? Does participation in restoration provide a foundation for something like "ecological citizenship" as well? This question may seem odd because the distinction between stewardship and citizenship may be unclear. The point, though, is actually more simple. Stewardship describes a kind of relationship between people and the land around them, but the Chicago restorationists also indicated that they were involved in a form of participation with one another as much as they were involved in meaningful participation with nature. If one of the goals of a democratic environmental theory is to not only work within the confines of our democratic institutions but also use environmental protection or restoration as a justification for strengthening those institutions, then one question would be, Can we expand the notion of participation in restoration and other environmental practices to consider it as part of the duties we might have to one another as members of a community? In short, can we understand such participation as a kind of civic obligation as well?

Ecological Citizenship

The goal of encouraging public participation in restorations has been previously characterized as representing a new and more expansive "culture of nature" (Light 2000a). Beyond producing a bond of interest between local communities and the nature around them, restorations also stimulate the development of moral norms more supportive of environmental sustainability in general. If restoration helps to produce such a culture of nature, though, what kind of culture will that be? Twentieth-century fascists arguably had a strong cultural attachment to nature that justified some of their most extreme and antidemocratic practices. A preferable culture context for our relationship with nature would be a democratic culture, meaning that the practices that would serve as a foundation for that culture should also be democratic. Ideally, participants in such a culture should see themselves as ecological citizens working simultaneously to restore nature and restore the participatory and strong democratic elements of their local communities.

What, though, is ecological citizenship? At first blush, it involves some set of moral and political rights and responsibilities among humans as well as between humans and nature. Although I do not have the space here to fully flesh out the appropriate contrasts, on this view, roughly, one's duties to nature ought not be isolated from one's duties to the larger human community. The goal of ecological citizenship would then minimally be to allow as many members of a community as possible to pursue their own private interests while also tempering these pursuits with attention to the environment around them. A strengthened relationship with nature promoted in this way would then entail the development of specific moral, and possibly legal, responsibilities or expectations that all of us be held responsible for the nature around our community and respect the environmental connections between communities.

Notions of citizenship in general, however, have a long history of philosophical and political debate and disagreement. Which understanding of citizenship would be best for infusing it with a set of environmental responsibilities as well? Although space prohibits a full explanation of the view, one useful understanding of citizenship for this discussion is along what is known as "classical republican" lines (not the political party), which identify a range of obligations that people have to one another for the sake of the larger community in which they live (see, for example, Pettit 1997). Thus, a duty of citizenship on this view is not satisfied merely by something like voting, and it is not exhausted by describing citizenship only as a legal category that one is either born into or to which one becomes naturalized. Instead, it is something that we might call an "ethical citizenship," or a concept of "citizenship as vocation," whereby being a good citizen is conceived as a virtue met by active participation at some level of public affairs. As the political theorist Richard Dagger puts it, what sets apart the "good citizen" on this view is that he or she does not "regard politics as a nuisance to be avoided or a spectacle to be witnessed" (Dagger 2000, 28).

The good citizen is someone who actively participates in public affairs, someone who generates "social capital" by active engagement with fellow citizens on issues of importance. Dagger and others are quick to admit that such an expanded sense of citizenship has been in steady decline throughout the history of the Western democracies. Citizenship is something that most of us today see as only a guarantor of certain rights but not as demanding responsibilities of us, other than leaving one another alone. Yet the language of citizenship still resonates widely in our culture as a way of talking about the moral responsibilities that people should have toward one another in a community. Defining what it means to be a "good citizen" is something that influential pundits outside the academy care about. Thus, using the language of citizenship to describe our relationship to one another and to the natural world could be a way of making discussion of such relationships more important to the broader public.

To add an environmental dimension to this expanded idea of citizenship would be to claim that the larger community to which the ethical citizen has obligations is inclusive of the local natural environment as well as other people. That is not to say that all legal citizens of a community would be required to become environmental advocates or ecological citizens in this way, but, rather, that embracing the ecological dimensions of citizenship would be one way of fulfilling one's larger obligations of this thicker conception of citizenship. In the same way, some people in our communities already join local parent-teacher associations as a way of fulfilling what they understand to be their personal and civic duties. Along these lines, contemporary republican theorists such as Dagger have already written much that helps us conceive of this kind of citizenship as inclusive of environmental concerns. Using the example of urban sprawl, Dagger (2003) argues that ethical citizens would have a good reason to fight sprawl because it threatens both the environmental and the civic fabric of a city. A sprawled city, as Whyte certainly appreciated, will only exacerbate the demise of civic associations that connect people to one another in networks of moral and political obligation.

If Dagger and others are right, then an expanded notion of citizenship is incomplete without an ecological dimension. And, if the point of ethical citizenship is to encourage people to take on responsibilities for one another in communities, then these responsibilities can be expanded to include environmental dimensions as well. If we look at things this way, then the volunteer restorationists in Chicago were acting as good ecological citizens in their participation in this set of projects. If those restoration projects were conducted only by contractors and did not involve public participation, then an opportunity to foster such ecological citizenship would have been lost. When people participate in a volunteer restoration, they are doing something good for their community both by helping deliver an ecosystem service and also by helping pull together the civic fabric of their home.

Another good example is New York City's Bronx River Alliance, a project of the City of New York Parks and Recreation Department and the nonprofit City Parks Foundation. The alliance is organized by a few city employees who coordinate sixty volunteer community groups, schools, and businesses in restoration projects along the twenty-three miles of the Bronx River. The focus is not only on the environmental priorities of the area; it is also on the opportunities to create concrete links between the communities along the river by giving them a common project on which to focus their civic priorities. Literature from the alliance says that the purpose of the project is to "Restore the Bronx River to a Healthy Community, Ecological, Economic and Recreational Resource." The alliance, like the Chicago restorations, is thus both civic and environmental, and the geographic scale of the environmental resource, crossing several political lines, helps create a common interest between them. Again, the project makes the environment the civic glue between various communities. (See Thalya Parrilla's essay in this volume.)

We must recognize, however, that the Bronx River Alliance did not emerge merely out of civic goodwill; it was formed by the City Parks Department in an attempt to follow other successful models such as the Central Park Conservancy, which has dramatically improved the ecological viability of Central Park while expanding citizen involvement in the maintenance of the park. The alliance was encouraged by the Parks Department leadership partly in response to funding shortages, which would have made it impossible to allocate sufficient public resources to restore the Bronx River without the work of the volunteers. But if we were to see public participation in such projects as an opportunity to restore, first, some bit of nature, second, the human relationship to that bit of nature, and third, the cohesiveness of the community itself, then creating the alliance would not be seen as a last resort under the conditions of budget shortfalls. Instead, it would be seen as the first choice for maximizing the various natural, moral, and social values embodied in this particular site. If we took the idea of encouraging ecological citizenship seriously, then we would want to create opportunities for people to engage in voluntary alliances of restoration (or other community environmental projects) even when we had public funding to instead pay parks workers or a landscape design firm to do the job for us.

The democratic participation of citizens in restoration projects is about building a democratic culture of nature or, more simply, a stronger human community that not only takes into account, but is actively inclusive of, concerns over the health, maintenance, and sustainability of larger natural systems. Such concerns will be important for the goal of encouraging the evolution of a more responsible citizenry overall, given the role such healthy environments play in making human communities themselves sustainable.

Recommendations

This discussion leads to two general recommendations for restoration based on the citizenship model. First, the expanded notion of ethical and ecological citizenship involves a robust notion of participation as direct democratic participation. Mere participation in an environmental project by allowing community input on an environmental decision is not enough, but it should be accompanied on this model by the creation of opportunities for people to actively engage in these projects on the ground. Such a framework is more likely to create a relationship between people and nature beyond mere stewardship, inclusive of seeing care for nature as a way of being a good citizen in their communities. Other hands-on environmental practices, such as community gardening, may also yield social values of citizenship equivalent to those of restoration (Light 2000a).

Second, along the lines of the citizenship model, the rights and obligations of people in an environmental community should be institutionalized. When something is designated as a right or responsibility under any understanding of citizenship, then it is eventually given legal status. If participation in democratic decision making is a right attached to citizenship, then we must have laws to ensure that citizens will be able to exercise their right to vote.

In the same way, if we took the idea of ecological citizenship seriously, then laws should be encouraged that mandate local participation in publicly funded restoration projects whenever possible. Because restorations become opportunities for forming bonds of citizenship they therefore take on the mantle of a state interest. The Bronx River example suggests the value of institutionalizing alliances between citizens and government. Another approach would be to mandate that democratically organized local citizen groups have a "right of first refusal" to participate in government-funded restoration programs. Thus, a restoration project request for proposals might stipulate that priority for license of the project will be given to voluntary organizations, subject to expert guidance. This requirement would resemble contracting provisions relating to local, minority, or women-owned contracting firms in government-funded housing projects. These regulations not only create local jobs, but also are intended to build local interest in such projects.

If government does not promote partnerships such as the Bronx River Alliance, then environmentalists should encourage such participation themselves. For instance, the Chicago Wilderness has involved the leadership of The Nature Conservancy, which has purchased land for restoration as well as coordinated volunteer restorationists on public lands. Likewise, the Field Museum in Chicago has donated office space for the coordination of these projects.

Larger restorations such as the multibillion-dollar project by the Army Corps of Engineers to restore the Florida Everglades may be too unwieldy for significant voluntary efforts, at least in terms of hands-on public participation. Smaller-scale restorations, such as the Chicago projects and Bronx River restorations, are ideal for this purpose, however. Although some environmental organizations favor larger, "wilderness"-oriented projects of preservation or restoration over such smaller-scale urban projects (Light 2001), we must, again following Whyte, narrow our geographic focus to consider the benefits of less flamboyant, smaller-scale initiatives in cities. More important, we must take from Whyte's earlier observations that the push toward more democratic participation in such projects will better serve the long-term interests of sustainability, conceived not as a narrow environmental goal, but as a more complete project that better connects local citizens with their local surroundings.

References

Curtin, D. 1999. *Chinnagounder's challenge: The question of ecological citizenship.* Bloomington and Indianapolis: Indiana University Press.

Dagger, R. 2000. Metropolis, memory, and citizenship. In *Democracy, citizenship, and the global city,* ed. E. F. Isin, 26–39. London: Routledge.

———. 2003. Stopping sprawl for the good of all: The case for civic environmentalism. *Journal of Social Philosophy* 34(1): 28–43.

Elliot, R. 1997. *Faking nature.* London: Routledge.

Gobster, P. H., and R. B. Hull, eds. 2000. *Restoring nature: Perspectives from social sciences and humanities.* Washington, DC: Island Press.

Gundersen, A. 1995. *The environmental promise of democratic deliberation.* Madison: University of Wisconsin Press.

Kant, Immanuel. 1785. *Grundlegung zur Metaphysik der Sitten.* Riga.

Light, A. 2000a. Ecological restoration and the culture of nature: A pragmatic perspective. In *Restoring nature: Perspectives from the social sciences and humanities,* ed. P. H. Gobster and R. B. Hull, 49–70. Washington, DC: Island Press.

———. 2000b. Elegy for a garden: Thoughts on an urban environmental ethic. *Philosophical Writings* 14:41–47.

———. 2000c. Restoration, the value of participation, and the risks of professionalization. In *Restoring Nature: Perspectives from the social sciences and humanities,* ed. P. H. Gobster and R. B. Hull, 163–181. Washington, DC: Island Press.

———. 2001. The urban blind spot in environmental Ethics. *Environmental Politics* 10(1): 7–35.

———. 2002. Contemporary environmental ethics: From metaethics to public philosophy. *Metaphilosophy* 33(4): 426–49.

Light, A., and A. de-Shalit, eds. 2003. *Moral and political reasoning in environmental practice.* Cambridge, MA: MIT Press.

Light A., and E. Higgs. 1996. The politics of ecological restoration. *Environmental Ethics* 18(3): 227–47.

Miles, I., W. C. Sullivan, and F. E. Kuo. 2000. Psychological Benefits of Volunteering for Restoration Projects. *Ecological Restoration* 18(4): 218–27.

Palmer, C. 2003. An overview of environmental ethics. In *Environmental ethics: An anthology,* ed. A. Light and H. Rolston III. Malden, MA: Blackwell Publishers.

Pettit, P. 1997. *Republicanism.* Oxford: Oxford University Press.

Pickett, S. T. A. 2003. Why is developing a broad understanding of urban ecosystems important to science and scientists? In *Understanding urban ecosystems: A new frontier for science and education,* ed. A. R. Berkowitz, C. H. Nilon, and K. S. Hollweg. New York: Springer.

Sagoff, M. 1988. *The economy of the earth.* Cambridge: Cambridge University Press.

Stevens, W. K. 1995. *Miracle under the oaks.* New York: Pocket Books.

Toth, L. A. 1993. The ecological basis of the Kissimmee River restoration plan. *Biological Sciences* 1:25–51.

Wald, M. 1999. White House to present $7.8 billion plan for Everglades. *New York Times,* 1 July, 14(A).

Wenz, P. 2001. *Environmental ethics today.* Oxford: Oxford University Press.

Whyte, W. H. 1968. *The last landscape.* New York: Doubleday. Republished Philadelphia: University of Pennsylvania Press, 2002.

PART FOUR

A More Humane Metropolis
for Whom?

As noted in the introduction to this book, early critics of "urban sprawl" like Holly Whyte, Jane Jacobs, and Ian McHarg failed to recognize the grave socio-economic inequity of white flight from cities to the urban fringe. Yes, the loss of farmland and scenic landscapes was disturbing and often unnecessary. But barely acknowledged—owing to the prevailing mind-set of that generation—was the grievous unfairness of federal tax laws, mortgage guarantees, highway programs, and local zoning laws, all contrived by "organization men" to insulate the prosperous white middle class from blacks and the poor (Platt 2004, ch. 6). Thanks to such landmark studies as *Sprawl City: Race, Politics, and Planning in Atlanta* (Bullard, Johnson, and Torres 2000), there is a growing realization that urban sprawl is intimately related to racial and economic separation within U.S. metropolitan areas and indeed may have been the intentional means to achieve such a polarization by class and race.

Part IV addresses this dimension of the "humane metropolis" through essays by leading members of the urban planning profession (Edward J. Blakely; Deborah E. Popper and Frank J. Popper) and a leader in the environmental justice movement (Carl Anthony). They are joined by a recent graduate of the University of Massachusetts Amherst, Thalya Parrilla, who summarizes efforts to restore a semblance of green and community pride to the South Bronx, based on her summer internship position there.

References

Bullard, R. D., G. S. Johnson, and A. O. Torres, eds. 2000. *Sprawl city: Race, politics, and planning in Atlanta.* Washington, DC: Island Press.

Platt, R. H. 2004. *Land use and society: Geography, law, and public policy,* rev. ed. Washington, DC: Island Press.

Race, Poverty, and the Humane Metropolis

Carl Anthony

The truth is, I hadn't thought much about William H. Whyte for almost a decade until Rutherford Platt came to my office to discuss a conference on the humane metropolis, celebrating Whyte's life and work. I explained to him that I have long been dismayed that most writers I had read on urban design seemed to have little understanding of the role that issues of race had played in the shaping of the nation's cities and land policies. I told him that I had been enthusiastic about the writings of Holly Whyte over the years. I did not, however, see how one could have a contemporary conference about the "humane metropolis" without considering issues of race and environmental justice as set forth brilliantly in the book *Sprawl City*, edited by Robert Bullard and others (2000). A review of Whyte's writings reveals that his work on environment and development from the mid 1950s until his death seemed to move him progressively closer to embracing the challenges of racial diversity. His insight about the importance of containing sprawl and reinvesting in cities helps lay the ground work for a new narrative that brings together the claims of racial and economic justice with those of ecological integrity as essential parts of the quest for a humane metropolis. To incorporate these claims fully, however, we need a larger framework than Holly Whyte developed.

In his influential book, *The Organization Man*, Whyte criticized the homogenizing influence of large corporations and other organizations on the quality of suburban life in the 1950s. He advocated more scope for individual initiative, both within the workplace and in suburban neighborhoods. He was also alarmed by suburban sprawl and advocated passionately for conservation of open space surrounding our metropolitan regions. In his 1957 essay "Urban Sprawl," he criticized the Interstate Highway Act of 1956 and its explicit intention "to disperse our factories, our stores and our people; in short to create a revolution in living habits." (quoted in LaFarge 2000, 132). He complained that affected communities have little to say about how the program, almost entirely in the hands of engineers, would be implemented.

The Exploding Metropolis (Editors of *Fortune* 1957) was perhaps the first book to raise concerns about postwar "urban sprawl." Essays by Whyte, Jane Jacobs, and others discussed suburban sprawl, transportation, city politics, open space, and the character and fabric of cities. "In this second decade of post war prosperity," Dan Seligman writes in one of the less-remembered essays, "in a time of steady advancing living standards, the slum problem of our great cities is worsening."

As noted in the introduction to this volume, Seligman also wrote the laconic statement, "The white urban culture they [poor nonwhites] might assimilate *into* is receding before them; it is drifting off into the suburbs" (Editors of *Fortune* 1957, 97). "Drifting off" is certainly a nonjudgmental way to describe the process of white flight in response to the *pull* of government incentives for suburban development and the reciprocal *push* of central city neglect. See also Ray Suarez, *The Old Neighborhood* (1999).

In *The Last Landscape* (1968), Whyte suggested in great detail a number of practical ways to conserve suburban open space, including the use of police powers, outright purchase, conservation easements, taxing policies, greenbelts, physiographic studies, cluster development, the design of play areas and small spaces, and scenic roadway design. He argued eloquently for increasing the density of urban and suburban communities to reduce costs, improve efficiency, and improve the quality of life of its residents.

In his magnum opus, *City: Rediscovering the Center* (1988), Whyte examined the social life of public plazas, streets, atriums, galleries, and courtyards, with detailed attention to what makes such spaces attractive or uninviting to the people who use them. He conducted detailed, empathetic investigations of the needs of street people, including vendors, street entertainers, people who hand out pamphlets, bag ladies, beggars, political activists, shopkeepers, postal carriers, and sanitation workers. In his observations and recommendations for improving the quality of street life, Whyte acknowledged those who are often left out of official planning consideration, which he termed "undesirables" (as deemed by society, not by him), by which he meant "winos, derelicts, people who talk out loud in buses, teenagers, and older people" (p. 156). Clearly, this new work was moving in the direction of helping city builders understand, acknowledge, and embrace the challenges of urban economic and racial diversity.

Although Holly White did not focus on race, his major works were written against the backdrop of an expanding consciousness about the importance of race in U.S. cities. I was seventeen years old in 1956 at the time that Whyte published *The Organization Man*. I lived in Philadelphia not far from Chester County, where Holly Whyte had grown up. The old road that connected Philadelphia to Chester, completely built up with residences, stores, and apartment buildings, was a block from our house. A trolley ran along Chester Avenue, and the street itself served as a sort of dividing line between our neighborhood, which was changing, and the all-white neighborhood on the other side. As the blacks from the South were moving into our neighborhood, the whites were moving out to the cookie-cutter suburbs that Holly White described in his case study of Park Forest, Illinois, in *The Organization Man*.

Whyte noted that the suburban community of Park Forest was an economic melting pot in the 1950s, a place for "the great broadening of the middle, and a sort of 'declassification' from the older criteria of family background" (Whyte 1956, 298). The suburbs, he observed were, compared with the residents' original communities, places of religious and social tolerance, provided one had the minimum economic wherewithal to rent or purchase. "This classlessness," Whyte notes, in the only paragraph I found about racial issues, "stops very sharply at the color line. Several years ago, there was an acrid controversy over the possible admission of Negroes. [For many Park Forest residents who] had just left Chicago wards which had just been 'taken over,' it was a return of a threat left behind. . . . But though no Negroes ever did move in, the damage was done. The issue had been brought up and the sheer fact that one had to talk about it made it impossible to maintain unblemished the ideal of egalitarianism so cherished" (Whyte 1956, 311).

Whyte stops short of speculating on the effect of this exclusion on the black families that were not allowed to join Park Forest. Nor does he develop the theme that huge public subsidies were beginning to support a new pattern of racial segregation in the metropolitan regions that by the end of the century were to become the dominant pattern of the nation. (The subsequent history of Park Forest, including its racial and commercial metamorphosis during and after the 1960s, is recounted in a recent video film by James Gilmore titled *Chronicle of an American Suburb*.)

I left home in 1956 and traveled through the American South. Separate drinking fountains and separate seating areas for colored and white were everywhere. Elvis Presley had hit the top of the charts, and Martin Luther King had not yet been elected president of the Montgomery Improvement Association, where he was to lead the bus boycott that made him famous.

If *The Organization Man* came to dominate some part of the national psyche in the 1950s, then the 1960s were dominated by its "shadow side": rejection of large organizations and male white chauvinism. By 1968, when *The Last Landscape* was published, the fury of the civil rights movement was reaching its peak. That was the year Martin Luther King was shot. America's metropolis was seething. Insurrections broke out in 168 cities. Rioting and looting claimed the lives of hundreds of people and resulted in billions of dollars of damage from Newark, New Jersey, to Los Angeles, California. It was the year the Kerner Commission reported that the United States was becoming two societies, one white and one black, separate and unequal. Whyte did not explicitly mention the theme of race in *The Last Landscape*, but by 1968, the dynamic of urban abandonment related to suburban sprawl was already well under way.

By the beginning of the 1990s, shortly after *City: Rediscovering the Center* appeared, the environmental movement in the United States had reached the peak of its influence, but most environmentalists were in denial about cities and race. On March 15, 1990, 150 civil rights organizations wrote a famous letter to ten of the largest environmental organizations, complaining that the environmental movement was racist. They pointed out that the memberships, staffs, and boards of these organizations included no people of color. Most important, was that environmental groups framed issues in a way that excluded and often went against the interests of communities of color. The disproportionate siting of hazardous waste facilities routinely placed in communities of color was ignored as an environmental issue by environmental groups (Sierra Club 1993).

Despite the advances in race relations during the previous four decades, environmental justice advocates pointed out, residential segregation based on race was more widespread than at any earlier time in U.S. history. The consequences of segregation had devastating effects on families in communities where more than 40 percent of the population lived below the poverty line. Employment opportunities were bleak. Education was poor. All these issues had environmental implications unnoticed by established environmental organizations.

Race, Ecology, and Cities

These issues call for a new narrative that integrates ecological awareness, issues of race, and patterns of metropolitan development. On one hand, advocates of ecological integrity must treat more systematically the concerns of social, economic, and racial justice in our metropolitan regions. On the other hand, proponents of social, economic, and racial justice must help build a shared understanding of the role of space, place, and ecological resources in the issues they care about (Bullard, Johnson, and Torres 2000).

For example, the conventional wisdom about people and nature in North America ignores the experience of communities of color. African American history illustrates why and how we must grapple with a more profound understanding of these relationships. Over the past several centuries, the ancestors of African American populations now living in cities have contributed to urban development and have been alienated from the natural world in many ways (Glave and Stoll 2006).

From the fifteenth century on, African American ancestors in Africa were brutally uprooted from a village context grounded in well-understood ways of life related to the stars and the seasons and adapted to climate, fauna, and flora. They were transported across the ocean and forced to work the land in North America, confined to rural plantations, without receiving the benefits of their labor. Although most blacks were kept away from the cities, the capital extracted and accumulated from their labor helped build the great world metropolises of Lisbon, Amsterdam, and London and, later, New York, Philadelphia, and Boston.

After the American Civil War, blacks were emancipated and promised enough land to be self sufficient: "forty acres and a mule." Although a few ex-slaves were able to re-create their traditional African cultures on the Sea Islands of Georgia and South Carolina, this promise never materialized for most. Instead, the federal government redistributed hundreds of millions of acres of land acquired from native people to railroad corporations and to new immigrants arriving from Europe. The majority of blacks, legally prevented from migrating to the cities, continued to work the land as sharecroppers and tenant farmers under a regime of state-sponsored terror. The wealth accumulated from their labor supported urban intermediaries in both northern and southern cities.

Finally, in the twentieth century, a combination of crop failure, mechanized agriculture, the boll weevil, and the lack of civil rights forced blacks off the land. Within a single generation a population, which for fifteen generations had been predominantly rural, became predominantly urban.

This journey of African Americans from rural areas to the cities is in many ways unique. Between 1940 and 1970, five million African Americans left the rural South for the urban North in the greatest mass migration in U.S. history. They left behind sharecropper shacks in Alabama, Georgia, Louisiana, Mississippi, and Texas for factory jobs and housing projects in Philadelphia, Baltimore, Detroit, Chicago, and Oakland, California. African Americans arrived in the cities en masse at the moment when the bottom was dropping out of the manufacturing economy. Middle-class whites were leaving in droves and taking their resources with them, abandoning the cities as a habitable environment.

In other ways, this migration of African Americans from rural areas to the cities is typical of people all over the world. The Irish, the Eastern European Jews and Catholics, Italian Americans, and Greeks were migrants who came through Ellis Island. Asian Americans, Pacific Islanders, Native Americans, and Latinos have also been recent migrants to the city. Indeed, the majority of the world's population has migrated from rural areas to the cities. This story tells us that if we wish to re-create a healthy relationship between people and the natural world, then we must pay attention to the similarities and differences in urban population groups and the continuing challenges of justice and immigration.

The New Metropolitan Agenda

Today, we are living through a remarkable time with unprecedented opportunities to reenvision the way we live in cities. I believe, however, that issues of race and poverty, social and environmental justice, must be central to the way we envision a truly humane metropolis, bringing together people and nature in the twenty-first-century city. In this new century, we need a new narrative that defines the claims of racial and economic justice and ecological integrity as essential parts of the quest for a humane metropolis. As I see it, the humane metropolis must be a process through which major urban settlements made up of multiple centers of cities, towns, and villages can be redesigned, rebuilt and reinhabited based on principles of compassion and consideration. We must have compassion and consideration for both the human and larger living community from which it draws sustenance. From this perspective, advocates of the humane metropolis must think not only about conservation issues, regional greenspaces, working landscapes, and urban gardens, but also about the challenges of poverty and racism, which are consequences of an uncaring, overly materialistic society.

During the past three or four decades, awareness of the metropolitan regions as a focal point for public policy, governmental corporation, physical planning, and economic strategies has been growing. Beginning with the early work of Holly White and his colleagues who wrote *The Exploding Metropolis*, an increasing number of environmentalists, urban planners, and activists have alerted the nation about the squandering of land and energy resources, the traffic congestion, and the pollution of air and water resulting from conventional suburban development practices.

In the 1998 elections, according to Myron Orfield (2002), 240 state and local ballot initiatives dealt with land use and growth, including coordinated comprehensive planning, state land trusts, and moratoriums on new growth. Voters approved more than 70 percent of these issues. In 1999, 107 of 139, measures, or about 75 percent, passed. In 2000, growth-related ballot initiatives numbered more than 550, and 72 percent were adopted. What is extraordinary is that private citizens in the most affluent sectors of society are going outside the normal decision-making process to implement controls on conventional land development practices (Orfield 2002).

In recent years, many businesses have expressed a renewed interest in metropolitan-level coordination and planning of land use and development. They are looking to find new ways to address traffic congestion, the jobs-housing balance, housing affordability, and workforce training. Typically, corporations use the language of competitiveness to argue for more effective patterns of metropolitan regional decision making. In 1993, the Congress of New Urbanism, made up of well-known designers and developers, was formed to curtail sprawl, redevelop vacant parcels in cities, provide housing for all, and plan for public transit. The group advocated pedestrian-friendly communities and creating healthy places to live and work.

Urban and suburban elected officials are beginning to see the connections between current patterns of metropolitan development and problems of discrimination, social isolation, environmental damage, and economic difference. State legislators, county supervisors, and suburban mayors are learning that the suburbs are not monolithic. They are beginning to see that issues of poverty and race are challenges in the older inner-ring suburbs built in the 1950s and 1960.

In this context, there is an extraordinary opportunity for advocates of social and racial justice, and advocates of ecological cities. To achieve a humane metropolis in the coming decades, the central cities and the older suburbs must be rebuilt. The emerging metropolitan agenda is an extraordinary opportunity for bringing together the claims of racial and economic justice with ecological concerns to support a humane metropolis for everyone (figure 1).

There are important lessons about the relationship between people and nature in this discussion. First, we must development the habit of seeing the cities in their larger ecological context. Just as the cities of the sixteenth, seventeenth, and eighteenth centuries were built through the exploitation of slave labor and degradation of land-based communities around the world, so also are our cities reshaping the hinterland. This effect goes beyond global warming and destruction of the rain forest to include the uprooting of traditional societies, causing mass migrations across the planet.

Second, issues of race and poverty are central to the construction of the humane metropolis. People of color have a long history with burdens of urban development. They have long suffered from urban geographic and institutional constraints imposed by racism.

Third, people of color have agency. Their energy and creativity can contribute to urban solutions, but this strength must be acknowledged. Just as Africans Americans historically escaped the plantation to create maroon societies, developed gardens within the confines of the plantation system, and created schools and churches for community survival and development after the Civil War, community development corporations, churches, and social movements within communities of color today have important roles to play in rebuilding the humane metropolis.

Finally, there is an old saying in the environmental field: "Everything is connected to everything else." The farms, the small towns, the suburbs, and inner cities are all connected (figure 2). The humane metropolis must find new ways to balance and reinforce qualities unique to each context. This job is a social, economic, and political task as well as an aesthetic one worthy of all our talents and creativity at the beginning of a new century.

Under the old narrative, we saw that European Americans conquered the North American continent. The natural world was seen as a vast and infinite resource that could be raided for more production and consumption. If there were problems with the cities, then we could pack up and leave, throw them away, build new ones, and "Devil takes the hindmost!" Knowledge was organized around the needs and experiences of the European American middle class. Anything outside these needs and experiences simply did not exist. In short, the world of Ozzie and Harriet was flat. If you ventured too far out, then you would fall off the edge.

Today, ecologists and others are feeling the pangs of guilt and remorse for destroying the ecological basis of life. Many people are beginning to believe that the universe is alive, and this insight has important implications for the ways we design, build, and inhabit or cities. *In the ecologist's story, however, people of color do*

Figure 1 Schmoozing in a downtown minipark in Madison, Wisconsin. (Photo by R. H. Platt.)

Figure 2 Cooperation in community regreening, New Haven, Connecticut. (Photo courtesy of Colleen Murphy-Dunning.)

not exist. Their experience, insight, and creativity are not acknowledged as a resource for addressing the challenges of our farms, cities, and suburbs.

At the beginning of this new century, we need a new narrative to bring together claims of racial and economic justice with those of ecological integrity as essential parts of the quest for a humane metropolis. Holly Whyte has made an important contribution to this new story. His studies of the organization man first alerted us to the negative effect of social homogeneity on the quality of suburban life. He outlined the social disintegration caused by the "exploding metropolis." He gave us tools to protect remaining, vulnerable suburban landscapes. He redirected our attention to the importance of rediscovering the center of our public life in the cities. If he did not deal explicitly and wholly with issues of race and poverty, then he helped lay the foundations for a new narrative into which solutions to these challenges can be incorporated.

As Thomas Berry once wrote in his remarkable book, *The Dream of the Earth* (1988, 123):

> It is all a question of story. We are in trouble just now because we do not have a good story. We are in between stories. The old story, the account of how the world came to be and how we fit into it, is no longer effective. Yet we have not learned the new story. Our traditional story of the universe sustained us for a long time. It shaped our emotional attitudes, provided us with life purposes, and energized action. It consecrated suffering and integrated knowledge. We awake in the morning and we know where we were. We could answer the questions of our children. We could identify crime, punish transgressors. Everything was taken care of because the story was there. It did not necessarily make people good, nor did it take away the pains and stupidities of life, or make for unfailing warmth in human association. It did provide a context in which life could function in a meaningful matter.

An agenda for the humane metropolis at the beginning of the new century must not only include the rivers and trees, wetlands, and working landscapes. It must also include the whole of the human community.

References

Berry, T. 1988. *The dream of the earth.* San Francisco: Sierra Club Books.

Bullard, R. D., G. S, Johnson, and A. O. Torres. 2000. *Sprawl city: Race, politics, and planning in Atlanta.* Washington, DC: Island Press.

Editors of *Fortune,* 1957. *The exploding metropolis.* New York: Doubleday.

Glave, D. D., and M. Stoll. 2006. *To love the wind and the rain: African Americans and environmental history.* Pittsburgh: University of Pittsburgh Press.

LaFarge, A., ed. 2000. *The essential William H. Whyte.* New York: Fordham University Press.

Orfield, M. 2002. *American metropolitics: The new suburban reality.* Washington, DC: Brookings Institution Press.

Sierra Club. 1993. A place at the table: A *Sierra* roundtable on race, justice, and the environment. *Sierra* (May–June): 51–58, 90–91.

Suarez, R. 1999. *The old neighborhood: What we lost in the great suburban migration, 1966–1999.* New York: Free Press.

Whyte, W. H. 1956. *The organization man.* New York: Simon and Schuster. Republished, Philadelphia: University of Pennsylvania Press, 2002.

———. 1968. *The last landscape.* New York: Doubleday. Republished, Philadelphia: University of Pennsylvania Press, 2002.

———. 1988. *City: Rediscovering the center.* New York: Doubleday.

Fortress America

Separate and Not Equal

Edward J. Blakely

> *It says "stay out" and it also says, "We are wealthy and you guys are not, and this gate shall establish the difference."* JAFFE 1992

> *What attracts people, most, it would appear, is other people . . . urban spaces are being designed, as though the opposite were true.* WHYTE 1978, 16

The ability to exclude is a new hallmark for the new public space in the United States. Fear created by a rising tide of immigrants and random violence ranging from the terrorist attacks of September 11, 2001, to the snipers in the suburbs of Washington, D.C., in 2002 has transformed public areas with an explosion of public space privatization.

Gated communities are clear indicators of the spatial division of the nation by race and class. In the 1960s, suburban exclusionary zoning to achieve this result was challenged and, to some degree, rejected through judicial or legislative open housing laws. De facto residential exclusivity has since been pursued through the private housing market, which has built hundreds of gated communities since the 1980s under the rubric of "security" from threats to homes and their inhabitants. These private enclaves, of course, may not explicitly be marketed as racist—racial restrictive covenants are unenforceable—but high prices and marketing practices ensure that they will largely be occupied by upper-middle-class whites.

William H. Whyte had a great deal to say about this emerging form of development that excludes rather than includes. What Whyte opposed was the design of space that reduces human interaction. The new fortress developments are aimed, at least on the surface, at reducing opportunity for social contact with strangers and even among neighbors. If there is little contact, then where is the social contract? If there is no social contract, then who will support the "public" needs of society, affordable housing, parks, health care, education, and so on?

Whyte emphasized in his studies that people may say they want to get away from other people, but their behavior indicates that their real desire is for quality human contact in open settings. "Urbanity," Whyte wrote speaking of community living, "is not something that can be lacquered on (*like a gate*); it is the quality produced by the concentration of diverse functions . . . the fundamental contradiction in the new town (*gated community*) concept of self containment" (Whyte 1968, 234).

Redefining the City as Walled Common

Gated communities are a new form of residential space with restricted access such that normally public spaces have been privatized. They are intentionally designed security communities with designated perimeters, usually walls or fences, and entrances controlled by gates and sometimes guards. They include both new suburban housing arrangements and older inner-city areas retrofitted with barricades and fences. They represent a different phenomenon than apartment or condominium buildings with security systems or doormen so familiar to Whyte who lived in Manhattan. There, a doorman precludes public access only to a lobby or hallways, common space within a building. Gated communities preclude public access to roads, sidewalks, parks, open space, playgrounds, in other words, to all resources that in earlier eras would have been open and accessible to all citizens of a locality. And, because these amenities are maintained privately through homeowner or condo fees, the willingness to support parallel facilities for the rest of the populace through taxes is accordingly diminished. As many as eight million Americans have already sought out this new refuge from the problems of urbanization, and their numbers are growing.[1]

Gated communities are proliferating, as are other elite forms of residential development like the resort developments, luxury retirement communities, and high-security subdivisions with which they overlap. Their rapid spread over the last several years results from a number of socio-demographic trends, especially the expansion of the size of the upper-middle-class with rising disposable income, combined with a rising tide of immigration and the threats of terrorism in public places.

Gates range from elaborate two-story guardhouses manned twenty-four hours a day to rollback wrought iron gates to simple electronic arms. Entrances are usually built with one lane for guests and visitors and a second lane for residents, who may open the gates with an electronic card, a punched-in code, or a remote control. Some gates with round-the-clock security require all cars to pass the guard, issuing identification stickers for residents' cars. Unmanned entrances have intercom systems, some with video monitors, for visitors asking for entrance clearance (figure 1).

All these security mechanisms are intended to do more than just deter crime: they also insulate residents from the common annoyances of city life like solicitors and canvassers, mischievous teenagers, and strangers of any kind, malicious or not. The gates provide sheltered common space, open space not penetrable by outsiders. Especially to the residents of upper-end gated communities, who can already afford to live in very low crime environments, the privacy and convenience that controlled access provides is of greater importance than protection from crime.

Gated communities in the United States go directly back to the era of the robber barons, when the very wealthy sealed themselves from the "hoi polloi." One of the earliest was the community of Tuxedo Park, built in 1885 behind gates and barbed wire an hour by train from New York. Tuxedo Park was designed with wooded lake views, an "admirable entrance," a community association, and a village outside the gates to house the servants and merchants to serve it (Stern 1981).

In the same period, private gated streets were built in St. Louis, Missouri, and other cities for the mansions of the rich. Later, during the twentieth century, more gated, fenced compounds were built by members of the East Coast and Hollywood aristocracies.

These early gated preserves were very different from the gated subdivisions of today. They were uncommon places for uncommon people. Now, however, the merely affluent, the top fifth of Americans, and even many of the middle class can also have barriers between themselves and the rest of us as a sign of arrival into a new separate—but never equal—American elite.

Gated communities remained rarities until the advent of the master planned retirement developments of the late 1960s and 1970s. Communities like Leisure World in Arizona, Maryland, and other states were the first places where average Americans could wall themselves off. Gates soon spread to resorts and country club communities, and then to middle-class suburban subdivisions. In the 1980s, upscale real estate speculation and the trend to conspicuous consumption saw the proliferation of gated communities built around golf courses, designed for exclusivity, prestige, and leisure. Gates became available in developments from mobile home parks to suburban single-family tracts to high-density townhouse developments. Gated communities have increased in number and extent dramatically since the early 1980s, becoming ubiquitous in many areas of the country (figure 2). Today, new towns are routinely built with gated villages, and there are even entire incorporated cities that feature guarded entrances.

These developments are descendants not just of a tradition of elite enclaves but of decades of suburban design and public land use policy. Whyte (1968) in *The Last Landscape* warned that this pattern was antihuman and antinature. Gates are firmly within the suburban tradition of street patterns and zoning designed to reduce the access of nonresidents and increase homogeneity. Gates enhance and harden the suburbanness of the suburbs, and they attempt to suburbanize the city. This suburbanization of the city is precisely what Whyte opposed in his essays on sprawl (LaFarge 2000).

From their earliest examples, the suburbs aimed to create a new version of the country estate of the landed gentry: a healthy, beautiful, protected preserve, far from the noise and bustle of the crowded cities. But demographic, social, and cultural changes permeate throughout society, and the suburbs are changing and diversifying. Suburban no longer automatically means safe, beautiful, or ideal. As the suburbs age and as they become more diverse, they are encountering problems

Figure 1 Access is restricted by electronic gates. (Photo courtesy Kathleen M. Lafferty/RMES.)

Figure 2 Another "waterfront" gated community under construction. (Photo courtesy Kathleen M. Lafferty/RMES.)

once thought of as exclusively "urban": crime, vandalism, disinvestment, and blight. Gated communities seek to counter these trends by maintaining the ambiance of exclusivity and safety the suburbs once promised. They exist not just to wall out crime or traffic or strangers, but also to lock in economic position. It is hoped that greater control over the neighborhood will mean greater stability in property values.

Gated communities are elite not just because of what they include, but also because of what they exclude: the public, strangers, and "undesirables" (Whyte's non–politically correct but gently ironic term). The result is privacy and control. Gated communities center on this ability to control the environment, in part because home buyers believe it will help protect their property values. Stability in the neighborhood comes from similarity in the makeup of the residents and in the houses in the development, and that is expected to mean stability in property values (interview with Curt Wellwood, Curt Wellwood Homes, Dallas, Texas, November 29, 1994). This consequence is the direct antithesis of the public space that Whyte promoted as the best vehicle to reduce crime and improve community life. Whyte saw public space as an essential ingredient in creating the interactions that promote and preserve community.

Social Security Behind the Walls

Through their homeowner associations and the codes, covenants, and restrictions built into the deeds, the new privatized communities are also able to control and exclude a vast range of down-market markers. From the highest-end developments down to those that are most modestly middle class, gated communities regulate out any possibility of activities and objects considered lower class, such as the unguarded open space, plazas, and parks that Whyte favored.

Status is important to most people, be they working class or affluent; the differences lie in what status symbols are most highly valued and especially which are accessible. According to the American Housing Survey (AHS), among those households that earn more than $161,481 a year, living in an exclusive neighborhood is considered a symbol of status or achievement by nearly half; among the very wealthy, who earn more than $400,000 a year, living in an exclusive neighborhood is important to nearly 60 percent (AHS 2002).

Those in the middle class—those just behind this large affluent class—are now more able to afford the symbols of status previously reserved for the very rich. The American middle class has expanded greatly since World War II. Household size has dropped across the board. Household net worth has more than doubled. This transformation is spectacular by any measure, and it has allowed a distinctively new set of economic behaviors to emerge. This new middle class has substantial buying power. The average disposable income is increasing from 1969 when only

8.3 percent of families with children had high disposable incomes to 15.5 percent in 1996 (U.S. Bureau of the Census 1997, 23–196). Owning a second home is also part of the new higher-income affluent lifestyle. In 1992, almost 35 percent of all middle class Americans considered having a vacation home an essential lifestyle feature, up 10 percent over the previous decade ("Four Income Families" 1995). While developers are finding it hard to build affordable homes for the average American, there is no dearth of market for resort properties. As a result of high levels of disposable income, moderately wealthy people in their early forties are able to afford very high status properties that they live in for only part of the year. And more and more often, both luxury developments and second home communities are gated.

Gated communities also cater to two large, new submarkets. New, active, affluent retirees make up the first class. Retirees are living longer and better than ever before. There are now more than twenty-five million retirees in the United States, and unlike the 1960s, most of these live independently. They are getting younger; the average retirement age is now only sixty. They are wealthier than in earlier decades as well. People over age sixty-five with incomes in excess of $40,000 climbed from only about 5 percent in 1970 to more than 15 percent in 1990, and those with incomes below $10,000 fell from 50 percent to less than 30 percent (Hull 1995).

"Flexecutives," the new well-paid, status-conscious, and mobile corporate executives, make up the second new submarket. Their numbers are increasing as telecommunications and new forms of corporate structure make smaller organizations the rule rather than the exception. The newest such developments are occurring outside major metropolitan areas. They are now in what were once relatively small towns and the outer edges of exurbia, and in places like Santa Fe, New Mexico, far from the crime and congestion of big-city regions. These communities must offer this new class of executives considerable charm and visual appeal with "unique architecture; culture; outdoor recreation opportunities; high quality, unusual retail facilities; and in particular restaurants—all in a small town setting" (Charles Lesser and Co. 1994). Many gated communities are designed especially for this elite group, providing the distinctive environment and exclusive image the buyers desire.

For developers too, gates are seen as an economic benefit. With their often elaborate guardhouses and entrance architecture, gates provide the crucial product differentiation and clear identity that is needed in crowded and competitive suburban new home markets. And, although there is no clear evidence that gates add a price premium, many builders report faster sales in gated communities, and quicker turnover means thousands in additional profits (Carlton 1990, 1:3).

The gate is part of the package of design and amenities that sell houses by selling a lifestyle image with which buyers wish to identify. As one developer of gated communities in Florida said: "Selling houses is showbiz. You go after the emotions. We don't go out and show a gate in the ad. But we try to imply and do it subtly. In our ad, we don't even show houses. We show a yacht. We show an emotion" (interview with Ami Tanel, Avatar Development Corporation, December 12, 1994). In part, it is the emotional response to race and class that these subtle images—codes—help to convey.

What It All Means

A gated community, with its controlled entrance and walled perimeter, is the very image of elite space. A gate means exclusivity, the foundation of what it means to be in an elite group. Although gated communities have traditionally been an option only for the economic elite, the very richest Americans, now they are part of the residential options of the merely affluent and even of the middle class, yet they retain the image of eliteness because they retain the function of exclusivity. The people inside gated communities may not be elite, but their developments share all the traditional markers of the communities of the elite. These communities place the Whyte notion of communal residential space on its head as his notion is transformed into a new mark of exclusivity.

The exclusionary ideal of gated communities arises from the status associated with social distance. Of course, social distance has long been a goal of our settlement patterns; after all, the suburbs were built on separation and segregation. The suburban pattern, which gates are meant to maintain and intensify, erected social and physical walls between communities, compartmentalizing residential space. That is not what Whyte intended. He favored the creation of space that pulled people together, not that created new wedges between them. As Rebecca Solnit and Susan Schwartzenberg say as if speaking for Whyte in *Hollow City* (2000, 75): "A city is a place where people have, as a rule less private space and fewer private amenities because they share public goods—public parks, libraries, streets, cafes, plazas, schools, transit—and in the course of sharing them become part of a community, become citizens."

Today, with a new set of problems pressing on our metropolitan areas, separation is still the solution to which Americans turn. In the suburbs, gates are the logical extension of the original suburban drive. In the city, gates and barricades are sometimes called "cul-de-sac-ization," a term that clearly reflects the design goal to create out of the existing urban grid a street pattern as close to suburbs, which Whyte would oppose. Gates and walls are an attempt to suburbanize our cities. Neighborhoods have always been able to exclude some potential residents through discrimination and housing costs. Now, gates and walls exclude not only "undesirable" new residents, but also casual passersby and the people from the neighborhood next door.

The exclusivity of these communities goes beyond questions of public access to their streets. They are yet another manifestation of the trend toward privatization of public services: the private provision of recreational facilities, open space and common space, security, infrastructure, even social services and schools. Gated communities are substituting for or augmenting public services with services provided by the homeowners' association. The same is true of all of the private-street subdivisions, which are now the dominant form of new residential development. In gated communities, however, this privatization is enhanced by the physical control of access to the development.

The trend toward privatized government and communities is part of the more general trend of fragmentation, and the resulting loss of connection and social contact is weakening the bonds of mutual responsibility and the social contract. The problem is that in gated communities and other privatized enclaves, the local community with which many residents identify is the one within the gates only. Their homeowners' association dues are like taxes; and the responsibility to their community, such as it is, ends at that gate. At a focus group with public officials in 1994, one city official in Plano, Texas, summed up his view of the attitude of the gated community residents in his town: " 'I took care of my responsibility, I'm safe in here, I've got my guard gate; I've paid my [homeowner association] dues, and I'm responsible for my streets. Therefore, I have no responsibility for the commonweal, because you take care of your own."

Residents of gated communities, like other people in cities and suburbs across the country, vary in the degree they personally feel the connections and duties of community within and outside their developments. The difference is that in gated communities—with their privatized streets, recreation, local governance, and security—residents have less need of the public realm outside their gates than those living in traditional open neighborhoods. If they choose to withdraw, there are fewer ties to break, less daily dependence on the greater community. *In making this choice, we descend the scale of democracy and do harm to our society and our aspirations as a nation. Holly Whyte would have been amazed and ashamed at this course of our democracy.*

Note

1. Definitive numbers on gated communities are now available from the U.S. Census. According to the American Housing Survey (AHS) conducted by the U.S. Census in 2001, more than seven million Americans live in gated or controlled-access communities. The West has the largest number of gated communities (11 percent of the total), with 6.8 percent in the South, 3.1 percent in the Northeast, and only 2.1 percent in the Midwest (AHS 2002). From surveys conducted by the Community Association Institute, the organization that represents community associations, the average number of units in community associations is 240 (2,995,200 units in gated communities). From 2000 census data, average household size is 2.45 people. Renters are the largest number of persons living behind gates and barricades, accounting for 67.9 percent of all residents. Whites are 80 percent of the population living in walled or controlled-access communities. Blacks are 12.5 percent and Hispanics 12.5 percent (AHS 2002).

References

AHS [American Housing Survey]. 2002. U.S. Bureau of the Census for the Department of Housing and Urban Development Research Information Service. Washington, D.C.

Carlton, J. 1989. Behind the gate: Walling off the neighborhood is a growing trend. *Los Angeles Times* (8 October).

Charles Lesser and Co. 1994. Flexecutives: Redefining the American dream. *Advisory* (Fall):1–8.

Four income families. 1995. *Money Magazine,* February: 154.

Hull, J. 1995. The state of the nation. *Time* (30 January).

Jaffe, I. 1992. Gated communities controversy in Los Angeles. *All Things Considered.* National Public Radio, 11 August. Interview with a member of Citizens Against Gated Enclaves.

LaFarge, A. 2000. *The essential William H. Whyte.* New York: Fordham University Press.

Solnit, R., and Schwartzenberg, S. 2001. *Hollow city: The seige of San Francisco and the crisis of American urbanism.* San Francisco: Verso Books.

Stern, R. A. M., ed. 1981. *The Anglo-American suburb.* Architectural Design Profile. New York: St. Martin's.

U.S. Bureau of the Census. 1997. *Household income of families with children classified as high income 1969 to 1996.* Washington, DC: Government Printing Office.

Whyte, W. H. 1968. *The last landscape.* New York: Doubleday.

———. 1978. *The social life of small urban places.* Washington, DC: The Conservation Foundation.

"The Organization Man" in the Twenty-first Century

An Urbanist View

Deborah E. Popper and Frank J. Popper

In his first great book, *The Organization Man,* William H. Whyte (1956) offered a new perspective on how post–World War II American society had redefined itself. Whyte's 1950s America had replaced the Protestant ethic of individualism and entrepreneurialism with a social ethic that stressed cooperation and management: the individual subsumed within the organization. It was the age of middle management, what Whyte thought of as the rank and file of leadership, whether corporate, governmental, church, or university. Those of us who grew up in the 1950s had *The Organization Man* seep into our consciousness well before we heard of Whyte the urbanist. It formed our ideas about conformity, resistance to it, and the meaning of being part of an organization. The book and its title gave many of us reason to disparage the security the organization promised; that was for others but not for us. The William H. Whyte of *City: Rediscovering the Center* (1988), his last great book, might seem an entirely different person. In the early work, he wrote of people in groups, of their social interactions within institutional structures. The latter was about how people behave in space, not institutions. In fact, it primarily focuses on people using space apart from institutions—the street or the plaza, for example. Seemingly so different, these two books reveal two sides of the Whyte coin, namely the focus on the individual in relation to surrounding context: social, organizational, and physical.

The first book sets out a social analysis and critique that still provides useful guideposts even as society has changed. In the 1950s, the city was still at the center of American life. The 1950 census of population was the last in which large U.S. cities—eastern, midwestern, and western—were still gaining population. By the 1960 census, New York City, Baltimore, Chicago, and Detroit were losing population, their centers thinning as their suburbs grew.

This evolving regional landscape of suburbia was created by and for "organization men" reflecting Whyte's social ethic in groupthink decisions on location, architecture, space allocation, and landscaping. The outward emblems of suburbia— office complexes, shopping malls, and residential subdivisions (often gated)—in turn reflected the organizations that commissioned them: public agencies, consulting firms, universities, and development corporations. This physical imprint of organization decision making on urban structure validates some of

Whyte's deepest concerns: the dominance of bureaucracy over the individual, scientism and its worship of statistics, and the rejection of "genius" or the idiosyncratic.

In the process, the old Protestant ethic choices of living and working in less pretentious (and more affordable) premises back in the core cities were left behind, literally and psychologically. The "old neighborhood," in Ray Suarez's phrase (1999), trickled down to the nonwhite, the ethnic, and the poor who still sought to improve their lot through start-up enterprises such as convenience stores, small repair shops, nail and hair services, and ethnic groceries and eateries. This process of spatial disaggregation of metropolitan America into underclass-dominated inner-city neighborhoods and organization-centered suburbia has been nurtured significantly by the shaping of tax laws, transportation decisions, and land use zoning by and for the benefit of the organization man society (Bullard 2000; Platt 2004, ch. 6; and also see Introduction to this volume by Rutherford Platt).

Although Whyte did not directly address this dimension of the shaping of the organization world, he did allude to race in at least one pungent paragraph of *The Organization Man:* "The classlessness [typical of Park Forest, Illinois, and its counterparts] stops very abruptly at the color line. Several years ago, there was an acrid controversy over the possible admission of Negroes. [For many Park Forest residents who] had just left Chicago wards which had just been 'taken over,' it was a return of a threat left behind. . . . But though no Negroes ever did move in, the damage was done. The issue had been brought up and the sheer fact that one had to talk about it made it impossible to maintain unblemished the ideal of egalitarianism so cherished" (Whyte 1956/2002, 311). (See also Carl Anthony's essay in this volume.)

The Social Ethic and the Suburb

The Organization Man evolved from a series of stories Whyte wrote for *Fortune*, where he began working in 1946. The book started to take shape with an article that appeared in June 1949: "The Class of '49," a commencement-season piece. Whyte used the opportunity to compare the aspirations of that year's crop of men graduating from college with those of his own class a decade earlier. After many interviews, he found a fundamental shift, a substantive redefinition of expectations and aspirations. America's individualist and entrepreneurial founding culture had given way to what he termed a social ethic. Well-educated, elite American men no longer aspired to start their own companies. Rather than outwardly engaging in competition, particularly with one another, they preferred to take the more secure route of belonging to an existing organization where conflict was muted and conformity to group norms rewarded with raises and promotions.

Whyte followed up that story by examining what happened once the graduating seniors entered their organizations. The relatively benign world of school carried over in the form of expanding numbers of training programs. Life after college was more a continuation than a break. Just as the college fraternity or the military had instilled a group mind-set, so had the new business world. Further reinforcing this mind-set was the availability of veterans' home ownership loan guarantees, which required evidence of a reliable future stream of income (but screened out most nonwhites even if they were veterans).

Beneath these changes in aspirations lay redefined group dynamics. In essence, the change meant that the organization's pressure on the individual would not be resisted because the individual no longer saw a need for resistance. Whyte found that those entering the world of work after World War II saw their interests as coincident with the group—the organization and their sections within it— rather than in opposition or irrelevant to it. Whyte was less concerned with the whats of the changes than with the whys, but certainly the experience of war and military organization was formative. During the war, the military had applied social science group management techniques extensively. In addition, many of the new management class came from blue-collar backgrounds, with group consciousness developed through unions. Whyte attributed three key beliefs to the social ethic, each reinforcing the group dynamic and managerial orientation: "[1] a belief in the group as the source of creativity; [2] a belief in 'belongingness' as the ultimate need of the individual; and [3] a belief in the application of science to achieve the belongingness" (Whyte 1956, 7).

Within American business, management increased in importance throughout the twentieth century as the scale of business increased. Whyte found that management had grown to be nearly an end in itself. The organization and its parts were to operate smoothly; conflict was viewed as undesirable, and when it arose it required skillful application of human relations to bring the workforce back into alignment. Training programs thus emphasized personnel skills, which were treated like a science. Personality tests were administered to ensure good matches. The individual, if properly placed, would not only thrive, but would also have a form for his life with and for the company. Whyte saw this shift to testing as intrusive and misguided. He doubted that the instruments could, in fact, find the best people for the job, but they would needlessly collect personal information, making the individual vulnerable. He also found that the test results that selected for leadership would have screened out most of the companies' actual leaders.

The social ethic might operate against genius, but it required larger organizations. The earlier Protestant ethic emphasized growth through competition, innovation, and cost containment; personal virtue lay in thrift and self-reliance. The organization under the social ethic also required growth to accommodate its expanding group, but drawing on a different range of attitudes and behaviors, it did so by expanding the consumer economy. The group substituted for self-reliance.

Company insurance programs took over for personal savings, for example, but then required the company to sell more to support the new programs. The organization needed a consumerist society more than it needed an innovative society, a bigger and bigger pie to support the increasing numbers of people being trained, doing the training, and overseeing the training. Applying some of the human relations to consumption could increase sales, generating increasingly sophisticated market analyses.

By accepting the organization's needs as his own, the organization man became footloose, ready to go wherever he was sent. He was less rooted in a specific place and lived in a series of sprawling new "insta-places": fungible, electrically equipped, train-oriented, school-centered, male-dominated white middle-class bedroom suburbs.

Whyte's famous chapter titled "The New Suburbia: Organization Man at Home" was the last and longest part of the book (1956, 267–404). In this extended case study, he gently probed the social mores of Park Forest, Illinois, as an archetypal postwar corporate suburb. Park Forest opened in 1948 as a brand-new planned community built on the prairie, with its own commercial town center and six hundred buildings offering more than three thousand dwelling units. It pioneered the townhouse living: low-density, multifamily housing set around grassy courtyards. Park Forest offered clean, neat, predictable, and affordable housing for Chicago's many new managers' families. Fairlington in Arlington, Virginia, a similar development, served a comparable purpose for the many government employees at the same stage in their careers.

Developments like these helped the organization man and his family move comfortably wherever the company required. They offered quick community and similar people, one more way in which life after college resembled life during it. Whyte's investigations drew on his love of mapping and his extraordinary ability—later so evident in his urban work—to show how people use space and how space shapes their interactions. Location in the court had an effect on one's social role; those at the center of the block, rather than those at the outer edge, set the tone. Whyte also tracked length of residence and the frequency of moves of Park Forest residents. Even if they stayed in Chicago, they tended to change houses as they moved up the corporate ladder, finding more spacious place-equivalents as the suburban housing market expanded and segmented along with the rest of the American economy and organization.

Unlike many of his readers, Whyte never derided the organization man and his suburbs. He did not describe the people as conformist, nor did he see their homes as ticky-tacky, filled with people trying to keeping up with their neighbors. He was wiser and more generous than that. He wrote: "There will be no strictures in this book against 'Mass Man'—a person the author has never met—nor will there be any strictures against ranch wagons, or television sets, or gray flannel suits . . . how important, really, are these uniformities to the central issue of individualism? We must not let the outward forms deceive us" (1956, 10–11).

He questioned instead America's turn to the social ethic, not that he idealized the Protestant one. He was concerned that the individual's interest was not the same as the group's and that the confusion of the two had harmful consequences. He knew that the organization needed less defense from the individual than the individual needed from the organization. Their interests could overlap on occasion—why would one work for a corporation with which one had no mutual interests?—but not always. In the merging of the two interests, the individual lost important elements of personhood and privacy.

Whyte ended *The Organization Man* with a reminder that organizational interests derive from individuals: "Whatever kind of future suburbia may foreshadow, it will show that at least we have the choices to make. The organization man is not in the grip of vast social forces about which it is impossible for him to do anything; the options are there, and with wisdom and foresight he can turn the future away from the dehumanized collective that so haunts our thoughts. He may not. But he can" (1956, 404).

Resisting a "dehumanizing collective" suffuses his subsequent work. Those "social forces" created beltways that broke up communities, developments that eliminated special secret places in the woods, and duller streets empty of people. He looked closely to understand what influenced people's choices of public spaces, how they expressed their preferences, and what might give them more of what they wanted. He provided the form for early open space legislation. He observed how people use plazas and found they *like* to move the chairs, to be in the sun or not, to face a friend or to be alone. Why not develop plazas, then, in which people can shape place to *their* liking rather than being forced into some theoretically preferable mold?

Beyond the Organization Man?

By the 1970s and 1980s, the organization man seemed to have dated. The new generation of college students became so knowing that they could hardly say the term *organization man* without a smirk, followed up with a shriek to at least include women, too. The security the Depression-era children needed had no appeal to the university graduates of the 1970s, and for the generation after them the organization's promise of security did not even register. Time between moves declined, and then length of tenure within the same organization fell. The median length of tenure for workers is four years (U.S. Bureau of Labor Statistics 2004), and for most of the period since the organization man, that figure declined. The median tenure at Fortune 700 companies for the chief executive officer is five years, two years less than in 1980 (Neff and Ogden 2001). Temporary workers increased in number; California's doubled between 1991 and 1996 (Bara 2001).

The Death of the Organization Man by Amanda Bennett (1990) reported on the demoralizing managerial layoffs of the 1980s that revealed the flimsiness behind the organization man's sense of belonging. David Brooks's more recent *Bobos in Paradise* (2000) finds his educated, affluent group, the "bourgeois bohemians" ("bobos") vaguely and superficially influenced by the organization man concept. They are aware of the term and generally adopt its antiorganizational sentiment, but are unlikely to have any deeper acquaintance with Whyte. Brooks himself finds Whyte's critique a bit soft and ambivalent, seeing him as a *Fortune* writer more than a social critic (Brooks 2000, 118–20).

In their interesting follow-up study to Whyte's book, sociologists Paul Leinberger and Bruce Tucker (1991) found that by the 1980s, the next generation's social ethic had mutated into a "self ethic." Loyalty to the organization by now had no base, a point that became painfully evident as corporate firings skyrocketed. Economic security and suburban life had created the "me generation." The prototypical suburb was no longer clustered Park Forest but decentered Irvine, California. The world had changed from daddy commuting to the city and mommy home with the kids to a much more commodified world—from playgroup to day care. Even the organization fathers by the end of their careers had learned the truth of Whyte's perception: that the organization and the individual are not symbiotic. The world changed and so did the organization, but Leinberger and Tucker thought Whyte's management philosophy had persisted, and so did Whyte (Beder 1999).

The mutation of the social ethic over time is evident in various ways: less loyalty, more horizontal management, more team meetings. In addition, many of the organization men are now women. Today, though, we are all organization people because organizations dominate our world more extensively and intensively than ever. Whyte captured a moment when people seemed to live in *an* organization, but today we move seamlessly, effortlessly *between* organizations, from one to another. Sociologist Jeffrey Pfeffer wrote in 1997: "We live in an organization world. Virtually all of us are born in an organization—a hospital—with our existence ratified by a state agency that issues a certificate documenting our birth," and it keeps on going from there. The shift from self-reliance Whyte found in *The Organization Man* resulted in corporations providing widely unrelated services, especially personal ones, preferably at a profit. As their domain expanded, the enterprise of understanding organizations became more the subject of business and management schools than sociologists (Pfeffer 1997, 3, 14), a change Whyte anticipated.

A 1999 *Fast Company* article on job satisfaction invoked Whyte to portray the difference between yesterday's and today's workers. Showing its dot.com boom moment, the article notes: "The organization man had to check his identity at the office door. People today . . . are demanding the right to display their integrity and the opportunity for self-expression." The article went on to say, however: "The new workplace also holds forth the promise of community—or some semblance thereof. Because we spend so much time at work, and because teamwork is a core organizational value, we expect to develop close bonds with our teammates. . . . The workplace provides a sort of home. Indeed, for better and for worse, work and life outside work blur into each other" (Fast Company 1999). Today's organization people imagine themselves different, but note how critical the group remains as a source of belonging and creativity. Note further how quickly the community dissipated as the dot.com bubble burst in 2001.

Whyte saw postwar education as moving from the liberal arts and sciences to the technical, from the fundamental to the applied or vocational (Whyte 1956, 85). He thought this shift suited organization men for problem solving, but neglected to develop their ability to decide which problems needed solving, fiddling while giving people the false sense that they were fixing.

Whyte wrote about how one seeks common ground in the group, tending toward a common denominator. The current economy is much more intent on flexibility—finding niches instead of the standardization of group conformity—yet that flexibility requires even further-ranging design conformity in some ways. We gave up loyalty to one organization for loyalty to many organizations as the overarching structure for our society. Whyte's organization man enlisted in organizations whose economic structures were about to reorganize. They had expanded to capture all the economies of scale, but the managerial change from the Protestant ethic to the social ethic paralleled the shift from economies of scale to economies of scope (Knox and Agnew 1998, 191–94). Economies of scale capture savings by streamlining ever-larger production of a clear and narrow range of goods. Economies of scope streamline by coordinating dispersed production and expanding range of offerings. The transition from the one to other created an organization world.

Whyte's organization man moved smoothly between places—from college to work, from jobs in one city to another, from one residential suburb to another. We have increased the facility with which we do so even more, moving in and out of organizations, shifting roles as we go from worker to consumer to evaluator. As the organization man moved from place to place, around him place-specific characteristics generalized, jobs increasingly located in the suburbs, and the feeling of belonging the organization man sought at work became more elusive. Even more elusive is escape from organizations.

Daniel Pink's book *Free Agent Nation* (2001) positions itself as anti–organization man. A former Gore vice presidential staffer, Pink left the White House to become self-employed and then urged everyone else to do the same. The advice has many problems, but for it to work at all requires that there be lots of interchangeable spaces available, places with telephones, faxes, and computer outlets. Such space can be rented briefly, possibly by the hour or the month, shared with others, acquired, and divested. The rooms are minimally decorated, painted in neutral colors, hung with inoffensive art, and show few signs of individuality. The economy is flexible, on the move, so for whom do you build? Everyone. The hotel conference space works ideally as such a one-size-fits-all place, whether in town, near the airport, or at the intersection of several highways in Joel Garreau's *Edge City* (1991). It is also an ideal place for the nonstop seminars on personal improvement that mix the social ethic with the self ethic.

Urbanism and the Organization Man

Among the problems Whyte wanted to fix were those observed in the landscape changes wrought by organization thinking (as codified in federal tax laws, local building and zoning codes, and investor preferences). The postwar remaking of the United States tore up the earth for new construction; highways, suburbs, roads, new commercial centers, and new office and factory buildings filled the countryside. It urbanized the region and sapped the city. The regional shopping centers served as the cores of the settlement, substantially altering the usual process of development. Gathering spaces became privately owned (Cohen 1996, 1053). The economy went through periods of boom and bust and reorganization, constantly reinventing itself.

The landscape is now dotted with interchangeable organizational spaces. The conference hotel is an iconic example. Others include the industrial park, the big-box store, the mall, and the gated community. Each is designed to be reliably predictable and controllable. Even elite shopping districts are predictable; Madison Avenue and Rodeo Drive vary less and less over time. Information and consumption are the growth portions of the economy, both with only mildly distinctive sectoral spaces, and center cities get remade as entertainment and tourism districts while other kinds of work shift to the outer edges (Hannigan 1998; Soja 2000). Organizations now are on campuses, and universities are more like businesses.

Universities are large organizations with major building programs where output is measured in pages produced and grants obtained. Many of us who thought we were avoiding or outsmarting the organization man by going into academic life have found ourselves working away in large technocracies. (At least job security for the tenured meets the organization man's expectations, one of the few surviving places where it does.) Applied or professional fields such as business, education, and public health have continued to grow in line with what Whyte saw in the 1950s, whereas the humanities continue to lose ground. Whyte noted that science had become as collaborative as the corporate world, and as costly. He expected that funding requirements would discourage the asking of unpopular or awkward questions. Over time, collaboration and funding have produced some extraordinary work, but some of Whyte's fears have been realized. For example, in the sciences, major biotechnology companies support much more work on genomes than on protecting ecosystems. Even as universities increasingly support specialized technical knowledge, they are increasingly interchangeable spaces, trying to attract everyone and anyone so that all parts of the plant are revenue streams. One strategy is to subcontract space to get rent, and thus many schools host Starbucks and Taco Bell. They contract vending machines exclusively to Coca-Cola or Pepsi, creating more organizational interpenetration.

Whyte saw leisure being eaten up by the organization: one was expected to work a lot. Yet consumption was essential to the growth of the economy. The Protestant ethic prized frugality, which helped profits by holding down costs, but was less aimed at increasing consumption. The social ethic needed to commodify leisure and thus add it to the corporate menu of sales. Increased consumer time pressures meant, however, that leisure had to become interchangeable, become easier to pick up and put down. The megaplex is everywhere and shows the same thing, sells the same snacks. We know the drill. Gyms are big, but from city to city they vary more by acceptable body images than by anything else. Many are national chains that define their income niches and then tailor a huge array of services appropriately (Epaminondas 2002).

Shopping malls, like schools, are major organizations. Their developers are certainly part of the management class about whom Whyte talked. Lizabeth Cohen writes that developers saw themselves as "participating in a rationalization of consumption and community no less significant than the way highways were improving transportation or tract developments were delivering mass housing" (Cohen 1996, 1055–56), the rationalizing of life and space that so worried Whyte. We shop as part of leisure and that shopping is smoother for its predictability. We know which way to turn when we walk into a new mall. Mall expansion actually peaked in 1978, when big-box stores emerged to provide another rationalized predictable shopping space, again each laid out in the same way, from city to city (Jackson 1996, 1120). George Ritzer (1993) echoes Whyte when he refers to the McDonaldization of society, the application of the bureaucratic approach to more and more segments of life, from fast food to hospitals, schools, and theme parks. Ritzer puts consumption at the center of his image. Others offer even darker visions. Experience eventually becomes so flat that even the human interaction and excitement of the mall pales. Eugene Halton describes the phenomenon as "brain suck," mass quantities of low-grade experience that eventually turn the individual inward. The only place to avoid a rationalized world is in time alone (Halton 2000).

The rationalizing, of course, segmented the market; it allowed the organization to creep into our lives by finding ways to derive profit from all aspects of our being. The elderly are increasingly housed, fed, and cared for by organizations. Assisted living is a growth market with both large corporations and small. Vacant lots at the outskirts of cities acquire new, overscaled Victorian or Georgian buildings from which their aged residents rarely venture. We know the organizational world endures because its landscape has spread so effectively. Since 1960, acreages at urban, suburban, and exurban densities have each more than doubled (Theobold 2001, 553). How to resist?

Organizations seek predictability through regulation, and that trait lives on. The social ethic provides protection from the ruthlessness of the entrepreneur. Environmental agencies and zoning boards rationalize use of space (at least in theory). The League of Women Voters of Park Forest can ally with, among other groups, government scientists to make rules that protect the group. Suburban zoning ensures minimum lot size, plenty of square footage (for residents), and decent plumbing and wiring. Shopping centers must have enough parking spaces. Rules intended to ensure better environments help control abuses, but they also produce more generic places. The language of regulation becomes increasingly organizational. We are all stakeholders and must meet to agree on our best interests; the group will find consensus.

Urbanist Lessons for the Organization Landscape

Whyte's resistance to the power of the organization was through modest methods: empirical observation, building from the ground up. In *The Organization Man,* Whyte was concerned about education's move away from the fundamental orientation of the liberal arts to the applied or vocational approach. He was impressed and depressed that business had the most majors, to such a degree that it set the dominant tone of the campus (Whyte 1956, 85). As the quantitative revolution was surging in the social sciences, he was skeptical whether its proponents would know what questions to ask and how to match their methods to the questions.

His main method, he insisted, was social *study,* not social science. Social science assumed that it could produce "an exact science of man" (1956, 217–30). Whyte called that utopian, in fact dystopian (Whyte 1956, 22–32) in the way that it intruded on privacy and limited human options. He delineates the difference in *The Organization Man*'s portrayal of management objectives, but the same problems arise in urban design, particularly when planning efforts attempt to root out the problems of the city through large-scale reimaginings that rationalize city life; contrast the mega-ideas of Le Corbusier with Whyte's more human-scaled ones. Whyte builds slowly and carefully, bit by bit, doing a form of market testing for public purposes. He advocates close observation of what works and of making what does not more like what does. In truth, utopian visions often seem to confer more reality on their imaginings, the future perfect, over the messy present (Donald 1999, 54). Whyte found more practical promise in the messy present. One could learn from the street vendors and bag ladies. They knew what places were popular (Whyte 1988, 25–55).

Despite his dislike of scientism, Whyte was systematic. He made painstaking counts, spent hours in thoughtful interviews, and devoted decades to looking at and noting activity on the city street. He measured angles and seat heights of benches in plazas. His counsel is clear. Use science: understand how watersheds work, how to maximize sunshine, select trees that will survive city stresses. (See the essay by Mary V. Rickel Pelletier in this volume.) Science can thus further social ends, but be careful in applying scientific method to social behavior. For that, use the softer social studies, and observe and think. Ask the right questions.

According to Paul Goldberger, Whyte saw the street as society's best achievement (LaFarge 2000, xviii). It is, in fact, the antithesis of the organization. Whereas the organization orders, provides hierarchy and structure, and determines the individual's behavior, street activity is driven by individuals; their purposes, their paths, and duration of stay derive from separate choices. People differ, so the best streets—and, by extension, the best societies—let them find their own ways. They vote with their feet, eyes, noses, and rears, choosing to walk and sit where they find something interesting. Let them take their best vantage points for observing the passing scene; encourage people-watching. Like his early *Fortune* colleague Jane Jacobs, his solution for high-crime areas was to attract *more* people, not fewer. Don't overdesign the park and the plaza; don't make the world too orderly. The more people watching, the more people-watching, the less crime. So let the activity commence with all types of people joining in. Privacy is also important, so ensure that as well, from the organization and from the street.

Whyte relied on time-tested strategies but relished challenging conventional wisdom that did not achieve desired goals. In *The Last Landscape,* for example, he noted that people thought that they were getting privacy when they bought a single-family home with a private lot (1968, 225–52). He found this belief a delusion—the design need not provide privacy just because it was a house in the middle of a lawn—and it was wasteful of land. He advocated "cluster housing" that included small, private gardens with wall heights that shielded one from random eyes. Thus, one did not need to spread across space to get privacy from one's neighbor; one merely needed thoughtful design. The idea of compactness in urban design conflicted with the prevailing trend toward urban sprawl, as reinforced by tax devices like accelerated depreciation and the interstate highway system (Hanchett 1996, 1082). Similarly, he suggested using another long-standing device, conservation easements, to maintain and attain open space.

Whyte also knew that after observing and making adaptations, one must keep monitoring and evaluating results. In *City: Rediscovering the Center,* he noted how planners used incentive zoning to good effect, getting developers to add some new public space. The plazas, he found, only sometimes worked well, however, and, even worse, newer ones were increasingly degraded through variances that neglected the original objectives. The device of incentive zoning was not sufficient; it had to be matched with consistent observation to see whether it continued to work as intended or needed adaptation yet again (Kayden 2000). (See the essay by Jerold S. Kayden in this volume.)

Land use and planning devices can preserve the sorts of landscapes and places that Whyte most valued: natural areas, the bucolic and productive countryside, and vibrant downtowns. Smart growth and New Urbanism owe some of their best insights to Whyte. Both aim to create walkable, mixed-use communities, heightening densities to generate liveliness while decreasing the environmental toll of more dispersed building patterns. They try to prevent fortress-style buildings and, where such structures arise, force an opening, a window, onto the street. Where corporate or commercial campuses get plopped down without a connection to neighbors, they try to build the connections. Perhaps their best hope is in the way in which they must work project by project. Smart growth and New Urbanism, for instance, derive from general principles that require that implementation include agonizing negotiations to make each project place-specific. Thus, new undertakings can be checked, altered, and improved, working from social study rather than social science.

The measure of all the urbanist devices, however, is the one Whyte hit on in 1956: the degree to which they support the individual. Organizational forces have enabled urban growth, turning cities into sprawling regions. Our days are bombarded by organizational obligations. Those trends may be unstoppable, but that should not stop us from trying to resist them. Whyte wrote: "It is wretched, dispiriting advice to hold before [the organization man] the dream that ideally there need be no conflict between him and society. There always is; there always will be. Ideology cannot wish it away; the peace of mind offered by the organization remains a surrender, and no less so for being offered in benevolence. That is the problem" (1956, 404). Our specific routes and routines should nurture us, however. If we keep in mind to make the kinds of spaces and places we enjoy, then we will always have some support for our own individual choices and protection from organizational imperatives.

Whyte's critique still holds today. He began and ended *The Organization Man* with the plea to let the individual shape the group rather than the other way around. Whyte gives us the forces that work on us every day: the convenience of organizations, their homogenizing of our world, and then our resistance as we try to keep space for our own eccentricities. His ongoing thinking about this conflict led him to advance ideas that became some of the most vital innovations in recent urban and environmental planning, things small and large like vest-pocket parks, downtown plazas, and smart growth. These innovations ensure that whatever role the organization plays in our individual lives, we have alternative spaces for our own particular enjoyment that can work as staging and supporting grounds for our own ideas and ideals.

References

Bara, S. 2001. Working on the margins: California's growing temporary workforce. Center for Policy Initiatives. Available online at http://www.onlinecpi.org.

Beder, S. 1999. Conformity not conducive to creativity, *Engineers Australia* (April): 60.

Bennett, A. 1990. *The death of the organization man.* New York: William Morrow.

Brooks, D. 2000. *Bobos in paradise: The new upper class and how they got there.* New York: Simon and Schuster.

Bullard, R. D., G. S. Johnson, and A. O. Torres. 2000. *Sprawl city: Race, politics, and planning in Atlanta.* Washington, DC: Island Press.

Cohen, L. 1996. From town center to shopping center: The reconfiguration of community marketplaces in postwar America. *American Historical Review* 101(4): 1050–81.

Donald, J. 1999. *Imagining the modern city.* Minneapolis: University of Minnesota Press.

Epaminondas, G. 2002. Find home, sweet home at the new haute gym. *New York Times,* 28 April, H:1–2.

Fast Company. 1999. Great Expectations. *Fast Company: The Magazine* (November): 212. Available online at www.fastcompany.com/online/29/greatexp.html.

Garreau, J. 1991. *Edge city: Life on the new frontier.* New York: Doubleday.

Halton, E. 2000. Brain-suck. In *New forms of consumption: Consumers, culture, and commodification,* ed. M. Gottdiener, 93–109. New York: Rowman and Littlefield.

Hanchett, T. W. 1996. U.S. tax policy and the shopping center boom of the 1950s and 1960s. *American Historical Review* 101(4): 1082–1110.

Hannigan, J. 1998. *Fantasy city: Pleasure and profit in the postmodern metropolis.* New York: Routledge.

Jackson, K. T. 1996. All the world's a mall: Reflections on the social and economic consequences of the American shopping center. *American Historical Review* 101(4): 1111–21.

Kayden, J. S. 2000. *Privately owned public spaces: The New York City experience.* New York: Wiley.

Knox, P., and J. Agnew. 1998. *The geography of the world economy: An introduction to economic geography.* 3rd ed. London: Arnold.

LaFarge, A., ed. 2000. *The essential William Whyte.* New York: Fordham University Press.

Leinberger, P., and B. Tucker. 1991. *The new individualists: The generation after the organization man.* New York: HarperCollins.

Neff, T., and D. Ogden. 2001. Anatomy of a CEO. *Chief Executive* 164 (February): 30–33.

Pfeffer, J. 1997. *New directions in organizational theory: Problems and prospects.* New York: Oxford University Press.

Pink, D. 2001. *Free agent nation: The future of working for yourself.* New York: Warner.

Platt, R. H. 2004. *Land use and society: Geography, law, and public policy.* Rev. ed. Washington, DC: Island Press.

Ritzer, G. 1993. *The McDonaldization of society.* Thousand Oaks, CA: Pine Forge.

Soja, E. W. 2000. *Postmetropolis: Critical studies of cities and regions.* Oxford: Blackwell.

Suarez, R. 1999. *The old neighborhood: What we lost in the great suburban migration, 1966–1999.* New York: Free Press.

Theobold, D. M. 2001. Land-use dynamics beyond the American urban fringe. *Geographical Review* 91(3): 544–64.

U.S. Bureau of Labor Statistics. 2004. Employee tenure in 2001. Available online at http://www.bls.gov/news.release/tenure.nr0.htm.

Whyte, W. H. 1956. *The organization man.* New York: Simon and Schuster.

———, ed. 1957. *The exploding metropolis.* New York: Doubleday.

———. 1968. *The last landscape.* New York: Doubleday.

———. 1988. *City: Rediscovering the center.* New York: Doubleday.

Sustainability Programs in the South Bronx

Thalya Parrilla

The South Bronx in New York City has the reputation of being a haven for drugs, prostitutes, and drag racing. Because of community intervention, however, its reputation is shifting. Today, a number of community programs are working to ameliorate social, economic, and environmental inequalities that are rampant in the South Bronx.

There are different definitions as to where the South Bronx begins. At the most southern part of the Bronx is Hunts Point. The neighborhood is generally broken into industrialized and residential areas. The industrialized area is concentrated around the waterfront, with most residential homes in the central area. High incidences of asthma, poverty, incarceration, and other social ills are found within this area. In response, a number of environmental justice organizations have sprouted from community members coming together over a cause or an injustice. Most of the organizations were formed in the mid-1990s and cover a wide range of issues and efforts. Many work in alliance with other organizations to remain strong and to gather support for their cause.

As the South Bronx continues to improve, there is concern that the neighborhood will be gentrified, thereby driving out the lower-income families and residents whose families are from the area. Gentrification is a reality in many neighborhoods throughout the five boroughs of New York City. The neighborhood improves, crime rates drop, and local residents no longer can afford the comforts of what they have been working to establish. The families that fought so hard to create a better quality of life are deprived of the fruits of their labor.

This essay concentrates mainly on Hunts Point, but includes organizations that also work within Port Morris and Mott Haven. Hunts Point, a peninsula that juts into the East River from the lower part of the Bronx, is squeezed between East Harlem, Port Morris, Rikers Island (a maximum-security prison), and industrial pollution. The population of Hunts Point is roughly ten thousand, mainly Latino and black; it has an unemployment rate of 24 percent.

The main truck access to the Hunts Point peninsula is a highway called the Bruckner/278. This highway, one of New York City's main arteries, funnels travelers and commercial traffic into the lower boroughs from southbound I-495. On a daily basis, some eleven thousand tractor-trailer trucks make their way to Hunts Point. The big rigs bring produce and meat to the markets at the tip of the peninsula surrounding the residential area. These markets are heralded as the largest in the world, and in 2005 the fish market in Manhattan moved to Hunts Point peninsula as well, further increasing the volume of traffic, noise, and carbon monoxide. Owing to the location of the meat markets at the tip of the peninsula, some of the rigs go off their designated routes and drive through residential areas despite complaints from residents concerned for the safety of the children.

Industries along the Hunts Point waterfront include a sludge processing plant, car repair warehouses, dump truck garages, transfer stations, junkyards, and truck repair shops. The workers generally speak only Spanish, and the owners of the industries live elsewhere and rarely go to Hunts Point.

The industrial area of Hunts Point is where most of the money is made. The money generally does not make it into the residential community because the trucks merely deliver their loads and go on their way.

A number of different organizations work out of Hunts Point. The neighborhood has become increasingly politicized about environmental justice issues and their link to the residents' health and quality of life. Many of the organizations are interested in educating the public and bringing more people into the organizations. The educational outreach component of these organizations is bilingual owing to the high concentration of Latinos living in the area. Many groups cater to the local youth to make the movement toward a better quality of life a sustainable effort as well as to get them involved in positive activities to stop the cycle of violence and poverty in the area.

Sustainable South Bronx

Sustainable South Bronx (SSB) is a community organization dedicated to supporting and implementing sustainable development projects for the residents of the South Bronx. The definition of "sustainable development" that SSB upholds is, according to its statement of purpose, development "that meets the needs and promotes an agreeable quality of life for the present without compromising the ability of future generations to meet their own needs and quality of life standards." The projects are based on feedback and information that address the needs of the community and uphold the values of environmental justice.

Majora Carter, a lifelong resident of Hunts Point, started SSB in 2001. Since its creation, SSB has won acclaim and recognition from various municipal entities and foundations for its success in implementing projects and supporting the values intrinsic to environmental justice. In 2005, Carter won a MacArthur Award for her work. Within the doors of the organization is a variety of bilingual informational literature that explains the principles of environmental justice, care of street trees, and other initiatives or events that are scheduled.

The core of SSB's effort is to address environmental racism that is present in the South Bronx. Studies have shown that the South Bronx has the least amount of accessible green open space in the city and in is in the top 10 percent for asthma in the United States. It is second to East Harlem in worst air quality in New York City. Some projects SSB has initiated are discussed below.

With a motto of "green the ghetto," SSB is working hard to reclaim blighted areas and convert them into green spaces for recreational use. One success story is the *Concrete Plant Park-in-Progress.* SSB, along with Youth Ministries for Peace and Justice, was integral in the renovation of ten acres of waterfront along the Bronx River, which now has been used for summer concerts and film screening. When completed, the Concrete Plant Park will be part of the vision of the South Bronx Greenway.

Not only does the lack of greenspace add to poor air quality, but the volume of traffic constantly exacerbates the problem. Because Hunts Point is a major hub for the fish, produce, and meat markets, it brings in approximately eleven thousand trucks daily. One effort to improve air quality in the neighborhood involved installing an *electronic truck bay.* This innovative technology allows a driver to plug the truck into an electronic console in the bay; heat, air conditioning, and other amenities are provided without the truck's engine idling, thereby reducing the amount of carbon dioxide in the air.

The Sheridan Expressway is an outdated and underused section of highway. SSB is working as part of the Southern Bronx River Watershed Alliance (which includes Youth Ministries for Peace and Justice, Mothers on the Move, and the Bronx River Alliance) to *decommission the Sheridan Expressway* and convert the twenty-eight acres it uses into a green open space, affordable housing, and other community needs. The project is still in its formative phase; working with outside sources, the Southern Bronx River Watershed Alliance is currently preparing a draft environmental impact statement.

Begun in the winter of 2003, the *Green the Ghetto Toxic Tour* takes participants to view twenty-three projects in progress and other sites that contribute to pollution in the area. The tour, with a guide to point out areas of interest, makes a number of stops, including at the Harlem River Rail Yards, the Waterfront (which harbors a number of industries), Concrete Plant Park, and the Hunts Point Riverside Park. Along the way are transfer stations, scrap yards, the prison barge, power turbines, a sludge pelletizing plant, factories, and a park.

SSB's Bronx Environmental Stewardship Training program, begun in 2002, provides hands-on training to participants in riverine and estuarine restoration as well as job readiness and life skills. The program, which runs for three months, certifies the trainees in OSHA regulations and trains entry-level tree climbers and New York City tree pruners. Other classes include various types of restoration, brownfield remediation, green roofs installation, wildlife identification, and hazardous waste cleanup. Upon successful completion of the program, participants can be hired by a network of employers. The participants are recruited from the community and work closely with a collection of organizations that are involved in the remediation and restoration of greenspaces in the Bronx and the other four boroughs. In the past, they have worked closely with youth groups and volunteers in the cleanup of the Bronx River. For instance, participants worked with Youth Ministries for Peace and Justice in one of their projects, Reclaiming Our Waterfront, or Project R.O.W.

A number of industries cut off access to the waterfront of Hunts Point, and some are derelict and have long since been out of business. Part of the plan for a greener Hunts Point is to revitalize the blighted areas in and around the peninsula. The *South Bronx Greenway* and *South Bronx Active Living Campaign* are two projects through which SSB is working to make that happen.

Across the Bronx River in Southview is a park with bike trails that has been built along the river. This area is important for restoration of natural habitat because the mouth of the river is home to a variety of wildlife. Hunts Point would like to replicate the effort and revitalize its side of the riverbank. The ultimate goal is to create a greenway that would allow people to use bicycles to get to Manhattan via Randalls Island (a recreational area that has swimming pools, baseball diamonds, and basketball courts). Currently, the only access to Randalls Island is through upper Manhattan by car or from the south by bicycle, thereby cutting off the Bronx, which is separated by six feet of water. The greenway would provide a space for recreation and contemplation in a parklike setting.

In coordination with SSB and the New York Economic Development Corporation, as well as a larger community visioning process, a private consultant and elected officials completed a master design, performed a feasibility study, and designed the greenway. There is strong interest in reclaiming the waterfront and making it accessible to families for recreation. Consistent with the trend in the rest of the United States, there is a high rate of obesity in the area. It has become a priority for urban areas to create recreational spaces for active exercise. (See the essay by Anne C. Lusk in this volume.)

The greenway project is part of a larger vision of developing an East Coast Greenway that runs from Florida to Maine. This concept is being gradually realized under the leadership of the East Coast Greenway Alliance.

SSB is working to creating green roofs in Hunts Point through the South Bronx *New Roof Demonstration Project.* In collaboration with Cool City Project at Columbia University, HM White Site Architects, and the Urban Planning Program, SSB is using the green roofs and public health research to create tangible results showing the economic and health benefits that stem from the use of green building technology.

The New York City Department of Sanitation, with funds provided by the New York City Council, coordinates free drop-off electronic recycling days in the five boroughs. The *electronic waste recycling project* was managed by INFORM, an organization involved with projects citywide. In the spring of 2003, SSB was charged with coordinating electronic waste recycling days within each of borough. SSB coordinated with Per Scholas, a company in Hunts Point that has onsite computer recycling, and Supreme Recycling, a computer recycling company out of New Jersey. The project was an overwhelming success. Electronic waste, by the tons, was saved from going to the landfill. The project, however, was a one-time venture for SSB. To date, the city funding for the borough-based waste coordinators that made it possible was for ten months. Groups are working throughout the city as part of the Zero Waste Campaign to re-create the conception in a bigger, better, and more sustainable way.

Community Gardens

Other groups have banded together to reclaim underused or blighted sites. Vacant lots are being claimed by community members and converted into community gardens. A number of organizations have been integral to the development of these green oases.

Green Thumb, Inc., established in 1978, now assists more than six hundred urban gardens throughout New York City (www.greenthumbnyc.org). Its central purpose is to nurture community participation in projects that contribute to neighborhood revitalization. This process involves acquiring derelict lots and transforming them into community spaces where members can grow anything from edible foods to flowers. The members have a direct influence on the design and function of the garden. Many of these gardens offer educational workshops for all ages, block parties to build community and membership, and food pantries.

Green Thumb offers its members technical assistance and materials in the form of soil, tools, and wood. Individual gardeners can also apply for grants to tailor their own plan and obtain the material required.

Green Guerillas started in 1973 in a Lower East Side garden in Manhattan. From there it has become a resource for different garden groups throughout the city. It provides support to the garden groups by helping with organization, planning, and outreach and in saving community space from further development. This group is an integral part to making a community garden a success. Specifically in the Bronx, Green Guerillas are involved with Trees for Life and Unity Project with La Familia Verde Coalition. This project was established to plant forty trees as a living memorial to the victims of the terrorist attacks of September 11, 2001.

Bronx Green-Up is an outreach program of the New York Botanical Garden that has similar functions to those of Green Thumb. Upon request, Bronx Green-Up provides informational workshops on a variety of technical training for members of community gardens. It also provides horticultural advice and hands-on assistance in garden maintenance. Bronx Green-Up has equipment and access to vehicles that support gardens in hauling garden supplies. At the core of the program are the community gardens and a regional compost education program.

The organization's close proximity to the Bronx Zoo affords it access to "zoo doo" to fertilize community gardens. The group also provides transplants for the gardens and gardening tools. All the services are free of charge.

Formed in 2000 by a group of concerned residents to address the epidemic rates of asthma within the community, *Greening for Breathing* (GFB) is an organization committed to supporting and increasing green infrastructure in the Hunts Point community. GFB's goal is to use trees as a green buffer from the industrial zone. A large part of this group's work is in supporting the small and growing population of trees. According to a 2002 tree survey conducted by the Parks Department, GFB, and volunteers, 67 percent of the trees are less than six inches in diameter. Most of the trees alive today have been planted within the last ten years as local citizens have sought to combat the high incidence of asthma.

GFB has worked with New York City Parks and Recreation, Teens for Neighborhood Trees (a group based in lower Manhattan that teaches tree stewardship), and other local groups. The organization has planted hundreds of young trees—street trees of approved varieties that do not trigger asthma or allergies—around Hunts Point.

With educational outreach the heart of its program, GFB produces bilingual newsletters and informational pamphlets on tree identification, care, and maintenance. It conducts training programs for members of the community, helping them become better informed and linking them to resources so that residents can become certified as "citizen pruners."

This effort is done in collaboration with the nonprofit organization *Trees New York*. This group is involved with different organizations throughout the five-borough area and has also begun an Adopt-a-Tree program with the New York Tree Trust.

In 2003, GFB worked with New York City Parks and Recreation to map the young trees, heritage trees, and possible sites for tree pits. Hundreds of possible sites were identified, and a proposal for the trees has been submitted. Enthusiastic efforts to implement the planting, care, and community outreach goals are under way.

Cooperatives

Other efforts taking place in and around Hunts Point include workers' cooperatives and river restoration. A fusion of the environment and art also benefits the community.

Green Worker Cooperatives (GWC) is an organization dedicated to worker-owned and environmentally friendly manufacturing businesses in the South Bronx. Started in the summer 2003 as an offshoot of New York City's environmental justice movement, GWC is involved in creating new alternatives to the working-class manufacturing jobs that have abandoned the South Bronx. With the flight of manufacturing companies has come a void that has been filled with high unemployment rates, low-paying service jobs, and polluting waste facilities. GWC is committed to addressing the economic and environmental issues. The organization looks for new ways to find gainful employment without harmful environmental side effects and exploitation of people. GWC conducts feasibility studies and business plans concerning advocacy, fund-raising, and recruiting. It seeks to convert bold ideas into a fully operating worker cooperative. After the inception of a cooperative, GWC provides training so that the workers can successfully realize their roles as the owners and their own advocates.

Solid waste management is a huge problem for all five boroughs of New York City. The waste issue has caused problems in many communities within the city, including the South Bronx. Creating industries to recover valuable materials in the South Bronx will serve to avert the need for more landfills and incinerators as well as preserve natural resources and reduce pollution.

According to the NYC Department of Design and Construction, the city generates thirteen thousand tons of "nonfill construction and demolition" waste per day, much of which is brought to the South Bronx. GWC's first project is the Building Materials ReUse Center and DeConstruction Service. Its objective is to recover building materials from construction projects that can be resold to supply low-cost building materials to projects in the area. GWC works in conjunction with ICA Group, a consulting firm with many years of experience in supporting worker-owned businesses.

With the redevelopment of the Bronx, action groups and environmental justice coalitions have worked to restore the riverbank area so that residents can enjoy the river. The Bronx River Alliance, established in 2001, is a consortium of public and private agencies to promote cleanup and public access to the river. It stems from a restoration community-based organization that was begun in 1974. One of its goals is to create a greenway that provides recreational access to the river, including canoe launch points and hiking trails. In some neighborhoods, abandoned factory sites are being reclaimed as greenspaces and community gardens. The Bronx River runs twenty-three miles through the Bronx; it is the only free-flowing river in the five boroughs of New York City. North of the Bronx in Westchester County, the river is accessible to local residents for recreation. Upon entering the Bronx, the river winds through Woodlawn Cemetery, Bronx Zoo, and New York Botanical Garden, where a surprising variety of wildlife and patches of old growth trees and natural land survive. After the New York Botanical Garden, the river is lined on either side by operating industries or by abandoned facilities. Its lowest segment before reaching the East River (an arm of Long Island Sound) is an estuary where fresh water from upstream mixes with tidal saltwater. Japanese knotweed, an exotic invasive that has choked out native flora, has invaded the riverbank along this stretch. This plant is difficult to eradicate because of its pervasiveness, the strength of its root system, and its ability to grow in disturbed areas.

The river is crisscrossed by four major highways serving New York City, and there have been cases of illegal dumping along the river by various industries. One such industry, a cement plant, dumped cement directly into the river. At the mouth of the river is a large sewage treatment plant that, during periods of high precipitation, overflows into the Bronx River along Soundview Park.

The Bronx River Alliance has an ecology team that consists of scientists and representatives from federal, state, local and city governments. According to the alliance's website (www.bronxriver.org), the team's ecological principle is "to minimize erosion, buffer sensitive natural areas, capture water on site, use recycled materials to the extent possible, maximize open space, include bird shelters, and educational signage." The alliance collaborates with local colleges and with the Army Corp of Engineers. For instance, Lehman College and the State University of New York's Maritime Institute are conducting a fish study and a watershed-wide soil survey in cooperation with the organization. The Bronx Zoo was identified as a major source of excess nitrogen leading to eutrophication of the river. The zoo has promised to aid in the cleanup of the river and to minimize animal waste discharge.

The goal of the restoration projects is to provide safe areas where families can recreate and enjoy a natural setting. The proposed greenway should also promote exercise to help combat obesity. In addition, it will serve as an outdoor educational resource for neighboring schools and potential outdoor enthusiasts while supporting the wildlife along the river. The Bronx River Alliance also coordinates river activities awareness activities such as canoe trips down the river and bike rides on the greenway in the North Bronx.

The alliance works closely with New York City Parks and Recreation as well as local environmental organizations such as Sustainable South Bronx and Youth Ministries for Peace and Justice. There has been solid support from other community groups and individuals, such as the New York City Environmental Justice Alliance, Congressman Jose Serrano, and the Wildlife Conservation Society. With the collective political pressure and a focused constituency, approximately $11 million has been secured to develop hiking and biking trails, to construct and restore wetlands, and to support projects to contain the overflow of sewage and stormwater. Since 2000, Bronx River improvement efforts have yielded forty acres of restored riverfront, 1.5 miles of greenway, three canoe launch sites, and removal of fifty derelict cars. Parties that are guilty of contaminating the Bronx River are being identified and held accountable for their actions.

It is intended that restoration projects will employ and train local community members to aid in the revitalization of the South Bronx by keeping the money within the community. The people involved in the river restoration projects are passionate about their work. There is hope that the main objectives can be met and an environmentally equitable future can be attained for the low-income area of the South Bronx. It is also hoped that the Bronx River can become a safe and healthy place for local residents to fish, learn about nature, and hike.

Another way local residents are approaching and interacting with the Bronx River is through art. The *Bronx River Art Center* (BRAC), funded by the New York State Council on the Arts, is a nonprofit organization created when the Bronx River Restoration Project began in 1980. It provides an artistic space for classes, exhibitions, and presentations. The Bronx River and the nature around it are brought in to be teaching tools and inspiration for artists. BRAC provides environmental programming for the community, including a parent-child team learning class on environmental studies and urban planning through the arts. The center also offers afterschool art sessions for children and youths, and evening adult classes. All classes are free and are bilingual.

Youth Programs

Other programs in the South Bronx aim to get youth involved. Through offering practical skills to developing leadership, certain organizations aim to increase the participation of young people in the area.

For instance, *Youth Ministries for Peace and Justice* (YMPG) is a nonprofit organization devoted to environmental justice issues. It seeks to foster peace within the community by involving community members, especially youth. Begun in 1994 by Alexie Torres-Fleming, it is involved in a variety of community projects, including the Bronx River Restoration Project. YMPG works primarily in the Bronx River, Bruckner, and Soundview neighborhoods in collaboration with United Way and the Bronx River Alliance. Its programs include reflection and study of peace, justice, and human rights through dance, mural painting, music, drama, sports, wellness, literacy development, journalism, photography, and video. Other programs focus on environmental justice, employment, education, community policing, and housing.

Project R.O.W. (Reclaiming Our Waterfront) consists of youths thirteen to twenty-one years of age who conduct monthly water monitoring for local and state agencies and take environmental educational canoe trips on the Bronx River. The group has also removed a number of derelict cars from the river, in cooperation with the National Guard. In addition, it has assisted in revitalization projects at Starlight Park and the Edgewater Road cement plant (a brownfield site) with funding from the National Oceanographic and Atmospheric Administration.

The Point Community Development Corporation has contributed to the cultural and economic revitalization of Hunts Point since 1994. This organization offers youth development classes, including hip-hop, environmental justice, environmental education, break dancing, and photography. It also hosts a number of festivals and holds theater productions put on by the youth of Hunts Point.

In one of its projects, ten young people from the community receive a stipend to create and implement a variety of initiatives to enact social change in the Hunts Point area. Some of these initiatives include the following:

- The Odor Journal to help residents report pollution
- Outreach project to address prostitution in the Hunts Point area
- Research on youth and community topics for monthly cable television presentations on BronxNet
- Creation of a weekly teen news and entertainment show on BronxNet
- Participation in the National People of Color Environmental Leadership Summit
- Research and dissemination of information to residents on proposals by businesses and city agencies that will affect Hunts Point.

Rocking the Boat engages youth in environmental education and in boat-building programs. Programs include such topics as traditional boat building, Bronx River habitat monitoring and restoration, community environmental education, forestry, what's next, and apprenticeships. These one-semester after-school programs for high school students target different subjects and aid students in realizing and pursuing goals after they graduate from the program. The apprenticeships are paid positions for students interested in continuing in the program.

All the programs work on creating practical skills that will help students in the future. The emphasis on the possibilities available in their urban and natural world provides a wider range of options for their future. The programs count toward high school credit. Rocking the Boat also holds several celebrations on the Bronx River to herald the launching of boats and canoeing for the community.

Recycling Initiatives

Among the different initiatives growing in the South Bronx is a new commitment to reduce the amount of waste that ends up in landfills. One program, *Per Scholas,* is a nonprofit organization that recycles computers and provides career opportunities by teaching computer technology classes. The recycling program began in earnest in 2000 when a state-of-the-art recycling facility was installed. Per Scholas's fifteen-week training course teaches local residents marketable computer technician skills. The core of the training includes extensive hands-on training by assembling, installing, and repairing the computers that are brought to the facility. Once the computers are restored, they are sold at a very low cost to underserved communities and schools.

The bulk of the computers that Per Scholas receives come from businesses and banks throughout New York City. In 2004, they partnered with Sustainable South Bronx for drop-off dates in communities throughout the five boroughs.

Another program, *Materials for the Arts,* collects surplus materials from companies and distributes them to organizations that need them. This program is responsible for removing tons of materials from the waste stream and recycling them.

Despite the reality of asthma, poverty, and drugs, the South Bronx has a rich and vibrant, community-supported culture. The reputation of the South Bronx as a drug haven has obscured the ingenuity of its population. Hunts Point and the surrounding neighborhoods have cultivated a community awareness to enrich the lives of their citizens. The nonprofit organizations that have sprung up in the homes of concerned community members are now established and have formed a strong alliance. The organizations mentioned are a cross section of many different efforts.

Because of the efforts of these organizations, today there are better ways for urban youth to spend their time, and jobs have been created to assuage the unemployment rate. Youth now have opportunities to be in nature and learn about the environment in a concrete jungle, learn how to use their environment to make art, and express themselves within that context. The counterbalance to all the positive developments is the relocation of the fish market to Hunts Point, and the traffic of trailer trucks has not decreased. The toxic industries still line the waterfront, although there are plans to remediate this area. Although some of the projected completion dates of the projects are years away, most projects are well under way, new and exciting projects have sprouted from their roots, and most of the funding has been secured. There is a hope and a drive in the organizations that root for a South Bronx that they envision. Their energy spurns on the movement for a greener and better Boogie Down Bronx.

PART FIVE

Designing a More Humane Metropolis

This book closes with a survey of issues and techniques for designing a more "humane metropolis" drawing on a variety of disciplines. In so doing, the discussion cycles back to Holly Whyte through reference to some of his practical contributions to the practice of urban design. The opening essay by Andrew G. Wiley-Schwartz of Project for Public Spaces, Inc. (a design consulting office that Whyte helped to found) recalls Whyte's "smile index" as a rough measure of a sense of well-being in shared urban spaces. (One can surmise that the "smile index" on today's freeways during rush hour falls below the chart!) Wiley-Schwartz goes on to identify three "threads" in Whyte's work: (1) sociability in urban space, (2) individuality afforded by cities, and (3) land conservation.

Next, Jerold S. Kayden, a professor of planning law at the Harvard Graduate School of Design, summarizes his study of "privately owned public spaces" in New York conducted on behalf of the city and the Municipal Art Society. His site-by-site survey of the design and management of more than public spaces procured through zoning incentives is reminiscent of Whyte's earlier work in both method and subject matter.

"Green urbanist" Mary V. Rickel Pelletier (a graduate of Kayden's school) explores some of Whyte's design principles as applied to site and building design, particularly the need to protect access to sunlight and daylight. She identifies various criteria for contemporary green building ("LEED-certified") that were anticipated in Whyte's writings.

Green architect Colin M. Cathcart draws on some of his own designs, both realized and hypothetical, to expand upon Rickel Pelletier's summary of green building criteria. As a resident of lower Manhattan with an office in Brooklyn Heights, Cathcart outdoes Whyte as a Big Apple–devotee by claiming—in common with a recent article in the *New Yorker* (Owen 2004)—that Manhattan is "greener" than the exurbs in terms of energy and time efficiency.

Finally, Timothy Beatley concludes the book, as he did the 2002 Humane Metropolis Symposium, with a review of the remarkable popularity and diversity of approaches to "green urbanism" found in most European cities today. Beatley argues that Europe offers an abundance of models for the United States. With the world just crossing the 50 percent urban threshold, these precedents need to spread quickly to the fast-growing megacities of Africa, Asia, and Latin America.

Reference

Owen, D. 2004. Green Manhattan. *New Yorker* (Oct. 18): 82–85, 120–23.

The Smile Index

Andrew G. Wiley-Schwartz

If there is a single symbol that sums up the work of William H. Whyte Jr., then it could be of the green bistro chairs scattered over the lawn of midtown Manhattan's Bryant Park. Whyte loved to watch people in a public park or plaza walk up to a movable chair, turn it an inch or two, and then sit down. The moves, he said, were important, not only allowing a person to express himself or herself in what is usually a proscribed environment, but also sending subtle social messages to those nearby. The message the chairs sent to the prospective sitter was even more important: this chair is here for you; do what you like with it. Like those movable chairs writ large, Whyte's ideas about people and cities respect the individual, focusing on how people relate to their surroundings and are able to express their individualism in it.

Bryant Park is Whyte's most enduring physical legacy. Known for decades as "Needle Park" owing to its pervasive drug trade, few New Yorkers entered it casually. Its main flaw was in its design: as a refuge from the city. To reinforce its separateness from the urban environment, the park sat several feet above street level, fenced off, and surrounded by high shrubbery. "The basic design . . . rested on a fallacy," said Whyte in *City: Rediscovering the Center* (1988, 160); "People say they want to get away from the city, avoid the hustle and bustle of people, and the like. But they do not. They stayed away from Bryant Park."

Whyte's prescription for Bryant Park, brought sharply into focus by the meticulous observational methods he developed using time-lapse film, interviews, and mapping in the park (conducted by Project for Public Spaces, a nonprofit group he helped launch), was simple: remove the walls and fences, let the people in and let them see in, give them something to do there—eat, watch movies, listen to music. Keep it clean. Let people decide where they want to sit. Renovations based on Whyte's recommendations have transformed that dangerous space into the grand, grass piazza that is Bryant Park today. Hundreds gather on summer weeknights to watch outdoor movies, and thousands use it every day. It is now New York's village green, its best small gathering place. The increase in property values that has resulted from Bryant Park's revitalization has been quantified at more than $20 million, but the intangible effect on the quality of life for New Yorkers is priceless.

It was Whyte's life's work to champion city life and demonstrate how we have continually acted to subvert it with poor planning and design. Whyte liked cities all through the 1940s, 1950s, and 1060s, decades during which cities experienced a dramatic loss of population and when many architects and planners were busy tearing down and redesigning their best attributes. Whyte's writings were a strong perspective shift for many readers. He saw city life as the sum of millions of marvelous interactions, each of which he sought to encourage. Businessmen returning from lunch linger at the top of a subway staircase, blocking traffic, deep in conversation. Lovers kiss, not in the shadows as one might expect, but in the most obvious place possible: right on the street corner. Lunchtime office workers defy convention, peel off their socks, and dip their feet into the pools at Seagram Plaza.

Ironically, it was at Seagram, that monument to internationalist style, where Whyte found his most successful plaza and the city found a reason to give incentives for plazas to other builders. Yet what Mies van der Rohe and Phillip Johnson had achieved by accident, other architects could not achieve even though they tried. Armed with the information and data he had collected observing the square day after day, Whyte concluded coldly, "It is difficult to design a place that will not attract people. What is remarkable is how often it's been accomplished."

Unlike the many critics who cast themselves as prophets but provide no solutions, Whyte searched for and gave answers. For more than sixteen years, as part of an ongoing investigation known as the "Street Life Project," he meticulously watched New York's public places, searching for the real reasons people gathered where they did. In one well-known study, the city asked him to study the bonus plazas of Sixth Avenue for signs of life. Finding few, Whyte offered deceptively simple prescriptive elements for their redesign. "This might not strike you as an intellectual bombshell," he liked to say, "but people like to sit where there are places for them to sit" (1980, 28). That was particularly true, he added, if they can watch other people from that vantage point. He saw that people love to sit near, touch, and play in water and that the heights of benches and walls frequently deterred people rather than accommodating them. To prove it, Whyte could rattle off ideal sidewalk widths, stair depths, street densities, and bench heights, and he could give specific, successful examples of each. In 1971, the city asked him to edit the planning code, using what he had begun to learn about urban behavior as a guide.

Whyte's scientific observations were coupled with a love for people and how they live and interact. He marveled at how people walk down a crowded street without bumping into one another. "The pace is set by New York's pedestrians and it is fast, now averaging about three hundred feet per minute. They are skillful, too, using hand and eye signals, feints and sidesteps to clear the track ahead. They are natural jay walkers, streaking across on the diagonal while tourists wait docilely on the corner for the light. It is the tourists, moreover, who are vexing, with their ambiguous moves and their maddeningly slow gait. They put New Yorkers off their game," he wrote in *The Social Life of Small Urban Spaces* (1980).

To Whyte, the vital city center was essential to a healthy civilization. It was as an editor of *Fortune* magazine in the 1950s that Whyte first called attention to the dangers the interstate highway system presented to city centers, the flight of young professionals from previously tight-knit ethnic neighborhoods, and the attempts to build the "city of the future" on the bulldozed remains of the vibrant city of the past.

Reading through Whyte's entire oeuvre reveals several threads winding through his work that indicate why he believed a vital urban core was so important. These thoughts are the underpinnings of his 1956 *The Organization Man* and the final conclusions of his 1988 *City: Rediscovering the Center.* The first thread is that the final revelation of Whyte's work observing and documenting what people did in cities convinced him that we are, despite everything we say, social beings who gravitate toward one another, when given half the chance. Or, as Whyte put it (1980, 19), "What attracts people most, it would appear, is other people." Seemingly obvious conclusions like this pepper Whyte's work, but so much of what we build continues to rest on the principle of isolation and separateness. When one considers that beliefs related to the deleterious effects of "crowding" and high densities (such as the commonly held negative correlation between population density and incidences of crime or disease) were used as justification for leveling whole neighborhoods, Whyte comes into focus as a crusader.

One little-known 1977 essay comparing street life in New York and Tokyo demonstrates his sharply reasoned but completely accessible style: "In the U.S., the conventional image of the high density core city is of a bad place, and bad not simply for its defects but for its essential qualities. . . . The image, unhappily, affects the reality it misrepresents; it is widely believed in Washington, not only by rural moralists, but by progressives who would save the city from itself. With few exceptions federal aid programs for cities have been laden with anti-density criteria which make it difficult for center city projects to qualify." To emphasize this point, Whyte toys with the idea of developing a "smile index" to prove that people are having a good time downtown:

> It is no frivolous matter, then, to note that many people on the streets of New York can be observed smiling, even laughing, and on the most crowded streets and at times, like the rush hours, when there might not seem much to be smiling about. New Yorkers themselves fervently deplore the city, its horrendous traffic jams, the noise and litter, the crowding. It is their favorite form of self-praise. Only the heroic, they imply, could cope. But they are often right in the middle of it all, and by choice; stopping to have a street corner chat, meeting people, arguing, making deals, watching the girls go by, eating, looking at the oddballs and the freaks.

Here, in this litany of center city attributes, is the second reason Whyte loved the city: because the very anonymity that it bestows on its inhabitants also encourages them to be individuals. To the writer who raged against conformity in *The Organization Man* and ridiculed corporate personality tests that filtered out potential employees with any personality to speak of, "freaks and oddballs" are what make city life interesting, especially in the form of street performers and vendors, whose cause Whyte routinely championed.

The third thread reveals itself in Whyte's prescient writings on land conservation. It is that Whyte, although he never says this explicitly, saw the built environment as a kind of ecosystem. His constant repetition that people do, in fact, want to be around other people, although they say differently, was a plea for rational development that protected open spaces near cities and towns. If city centers could remain vital places, then people would remain anxious to live in them, releasing the pressures on our open spaces that sprawl continually attacks.

As early as the 1950s Whyte wrote on land use preservation and open space protection, describing and debating the merits of cluster development, conservation easements, urban sprawl (it is possible he introduced these terms to the public), and other still current issues. Then, the dangers that the interstate highway system presented to city centers and the flight of white middle-class families were new, shocking concepts, and current beliefs and legislation on protecting open spaces and controlling sprawl are just catching up to them.

Ironically, Whyte's pleas for people-friendly environments were heard best by commercial developers with little or no interest in keeping urban neighborhoods or small towns intact. At any mall, one can see evidence of his prescriptions succeeding wildly. Unfortunately, malls are controlled, soulless environments, and they are in the suburbs, negatively charging the natural magnetism of the city center. Although Whyte had many offers to help guide the developments of malls and other private commercial places, he always turned them down.

Whyte understood that architecture and design were important, but he placed the people who had to live in the places they designed at the forefront of his analysis. These considerations, however, are still ignored by many architects and planners who are more concerned with the "statements" their buildings and public spaces make than the people who must live and work around them. "Architects and planners like a blank slate," Whyte wrote in *The Social Life of Small Urban Spaces* (1980). "They usually do their best work, however, when they don't have one. When they have to work with impossible lot lines and bits and pieces of space, beloved old eyesores, irrational street layouts, and other such constraints, they frequently produce the best of their new designs—and the most neighborly."

With the rise of new urbanism and center city projects such as James Rouse's festival marketplaces in Boston and Baltimore, one might suppose that Whyte's ideas are now widely held. Yet we must look closely at these new downtown spaces to see if they are performing to their full potential. Most, it must be admitted, still fail the test. The new ballparks are expensive and are only in use for brief periods of time. Boston's Quincy Market attracts tourists, but residents still shop at the Haymarket. So it is no stretch to see that the Organization Man would be happy in Celebration, Florida, which is essentially a company town peopled with Disney employees. We know that there is something "wrong" with places like Seaside, Florida, a town planned around New Urbanist principles, but it is hard to put a finger on exactly what that thing is. If he were here today, William H. Whyte could do it for us. Instead, we must vigorously apply his litmus test of everyday, diverse use to all these new suburban and urban developments; otherwise, we will wind up with temporary places, places without the essential friendliness and easy qualities that allow us to endow them with our memories by encouraging and facilitating chance meetings, important milestones, weekday lunches, and weekend festivals. It is the accumulation of all these types of events that are, to paraphrase Jane Jacobs, the small change upon which our cultural wealth is built.

References

Whyte, W. H., Jr. 1956. *The organization man.* New York: Simon and Schuseter. Republished Philadelphia: University of Pennsylvania Press, 2002.

———. 1980. *The social life of small urban spaces.* Washington, DC: The Conservation Foundation.

———. 1988. *City: Rediscovering the center.* New York: Doubleday.

Zoning Incentives to Create Public Spaces

Lessons from New York City

Jerold S. Kayden

In 1961, the City of New York inaugurated a new concept of "privately owned public space" to be created by developers in exchange for zoning concessions.[1] Through a legal innovation known as "incentive zoning," the city granted floor area and height bonuses and other zoning concessions to office and residential developers who would agree to provide public spaces in the forms of plazas, arcades, atria, or other forms of indoor or outdoor space on their premises. Ownership of the space would remain with the developer and subsequent owners of the property, and access and use would be open to the general public, hence the term "privately owned public space." Cities across the country followed New York City's lead, encouraging their own contributions to this distinct category of urban space; see Lassar 1982, 17–18 (for Hartford, Seattle); Svirsky 1970, 139–58 (for San Francisco); and Getzels and Jaffe 1988.

How has this legally promoted marriage of private ownership and public use fared? This essay discusses the results of a three-and-a-half-year empirical study conducted by this author in collaboration with the New York City Department of City Planning and the Municipal Art Society of New York. The findings are fully reported in *Privately Owned Public Space: The New York City Experience* (Kayden 2000).

Most broadly, the study found that zoning incentives have had a considerable effect on the design of the city's ground plane, particularly by encouraging interposition of public spaces adjacent to or inside new buildings at the developer's expense. More specifically, the study found that although New York City's law yielded an impressive *quantity* of public space—503 spaces at 320 office, residential, and institutional buildings—it failed to deliver a similarly impressive *quality* of public space in terms of both initial design and subsequent operation. At their best, the spaces have combined aesthetics and functionality, creating superior physical and social environments, set intelligently within their surroundings. Members of the public use the best spaces for social, cultural, and recreational experiences. At their worst, the spaces have been hostile to public use. Many are nothing more than hapless grass strips or expanses of barren pavement, while others are privatized by locked gates, usurpation by adjacent private uses, and diminution of required amenities, in violation of applicable legal requirements.

This essay first explains the legal framework responsible for creating privately owned public spaces in New York City. It next describes the principal findings of the empirical study. Finally, it proposes changes to the responsible legal and institutional regime likely to promote improvements in the quality of privately owned public spaces in New York City and elsewhere.

Legal Framework

Privately owned public space is a legal oxymoron. "Privately owned" refers to the legal status of the land, the building, or both where the public space is located. The nature of the space's "publicness" is legally determined by the city's zoning and related implementing legal actions. The zoning law establishes the framework within which developers and designers exercise their creative abilities. Specified design standards have incorporated diverse visions of public space held by urban planners and designers, civic organizations, and public officials as well as by developers, owners, and members of the public. The applicable law is amazingly detailed on some aspects and remarkably terse on others. The design standards have changed over time, reflecting an evolution in thinking about what makes public space succeed or fail and how demanding and precise legal standards need to be to secure good outcomes.

Since 1961, the Zoning Resolution has defined twelve discrete legal types of privately owned public space, including *plazas, arcades, urban plazas, residential plazas, sidewalk widenings, open-air concourses, covered pedestrian spaces, through-block arcades, through-block connections, through-block galleries, elevated plazas, and sunken plazas.* In addition, the zoning has enumerated spaces that are geographically tailored to specific needs within special-purpose zoning districts. Regulatory flexibility allows "customized" public spaces not otherwise described in the Zoning Resolution to be accepted as a condition of development approval.

Although the level of detail and clarity vary greatly, the zoning provisions governing each public space type have specified (1) design standards, (2) the legal approval process, (3) the responsibilities of owners, and (4) the rights of members of the public to use the space. Sometimes the provisions have established mechanisms of enforcement to encourage owner compliance with the law. A three-tier set of legal actions under which spaces may be approved comprise (1) discretionary special permits and authorizations, (2) ministerial "as-of-right" approvals, and (3) an intermediate option called "certification." The applicable level of review depends on the cost and magnitude of the proposed project. To grasp fully the "law" for a given space, it is necessary to scrutinize the relevant Zoning Resolution provisions as well as the conditions of approval for specific sites.

To obtain more than five hundred privately owned public spaces, the city principally has relied on a voluntary approach known as incentive zoning. This approach offers a private developer the right to construct a building larger or different from what is otherwise permitted by the zoning; in return, the developer provides a privately owned public space.[2] The social rationale for this exchange is that the public is better off in a physical environment replete with public spaces and bigger buildings than in one with fewer public spaces and smaller buildings.[3] Essential to this approach are the assumptions that the zoning code is rigorously enforced and that variances of height and floor area are not otherwise obtainable.

Redolent of *Nollan v. California Coastal Commission*[4] and *Dolan v. City of Tigard*,[5] the legal rationale is that public space is "density mitigating" in that it counteracts the negative effects, such as street and sidewalk congestion and loss of light and air, potentially caused by larger buildings.[6] For the developer, the rationale is pure real estate economics: when the value of the incentive equals or exceeds the cost of providing the public space, the transaction becomes financially attractive.

The Zoning Resolution announces the nature and extent of the incentive for each type of public space. The primary incentive has been the floor area bonus, usually measured in relation to one square foot of provided public space. For example, a developer may receive a floor area bonus of ten square feet for every square foot of plaza, so a five-thousand-square-foot plaza would generate an extra fifty-thousand square feet of buildable zoning floor area.[7] Although the bonus multiplier for the different types of public space ranges from three to fourteen bonus square feet for every square foot of public space, proposed developments have always been subject to a bonus cap limiting the total bonus floor area earned from all provided public space to a percentage, usually 20 percent, of the base maximum zoning floor area. In zoning terminology, the bonus is an increased "floor area ratio" compared with what the zoning law would otherwise allow.[8] For developments on large lots, the Zoning Resolution has also authorized the use of non-floor-area incentives, such as waivers of applicable regulations affecting the height and setback of a building or how much of the lot the tower portion covers, to encourage the provision of public space.

The metrics of incentives are conceptually straightforward. To attract developers, incentives must convey a financial benefit exceeding the cost incurred in providing the privately owned public space. Zoning incentives benefit developers either by increasing income or reducing overall building cost. For example, a floor area bonus increases a building's cash flow or value through rental or sale of the extra space. Frequently, the ability to develop extra space allows the building to be taller, and the higher-story floors may be rented or sold at premium rates. Height, setback, and tower coverage rule waivers may allow a building design that is more in keeping with the tastes of the market or may decrease construction costs.

In return for the incentive, the developer agrees to allocate a portion of the lot or building for public use, to construct and maintain the space, and thereafter to allow

Figure 1   Entrance to Trump Tower and public space, New York City. (Photo by R. H. Platt.)

access use of the space by the public (figure 1). In effect, the developer "pays" for its bonus floor area or non-floor-area incentive by agreeing to these obligations. Although the space continues, by definition, to be "privately owned," the owner has legally yielded certain rights, notably the right to exclude the public from the designated space. The space is thus effectively subject to an irrevocable easement of public access.

Quantitative Results

In return for more than sixteen million square feet of bonus floor area,[9] the city obtained 503 privately owned public spaces at 320 commercial, residential, and institutional buildings. Categorized by the twelve legal typologies enumerated in the Zoning Resolution, the public space inventory includes 167 plazas, 88 arcades, 57 residential plazas, 32 urban plazas, 15 covered pedestrian spaces, 12 sidewalk widenings, 9 through-block arcades, 8 through-block connections, 3 through-block gallerias, 1 elevated plaza, 1 open-air concourse and 110 other spaces located in special zoning districts or uniquely defined by other legal means.[10] Not surprisingly, the production of public space corresponded with cycles of real estate development that flourished from 1968 to 1974 and from 1982 to 1989. The total area of privately owned public spaces was 3,584,034 square feet, or slightly more than eighty-two acres. To put this number in perspective, New York City's privately owned public spaces would cover almost 10 percent of Central Park, or thirty average Manhattan blocks.[11]

The geographic distribution of spaces established under the New York City density bonus incentive is overwhelming skewed to Manhattan and particularly its highest value real estate districts. Of the 320 buildings with such public space, 316 are situated in Manhattan, three in Brooklyn, one in Queens, and none in the Bronx or Staten Island. Within Manhattan, most public spaces are clustered in four areas: the financial district, midtown, and the Upper East and West Sides. This pattern of spatial concentration is simply due to the influence of the real estate market. By definition, the bonus yields public spaces only where developers want to construct buildings larger than allowed by the existing zoning. In general, high-rise, high-density districts with strong demand for additional floor area will be the loci for zoning-generated public spaces, whereas low-rise neighborhoods lacking such demand will not.

Geographical clustering within high-density areas makes public policy sense. Privately owned public spaces work best in crowded commercial and residential districts. In older, lower-density neighborhoods, where private yards are more plentiful, the kinds of spaces under consideration may offer less benefit. Furthermore, residents of such neighborhoods may oppose the very scale of development necessary to generate public space under incentive zoning. The lack of a geographically equitable distribution of usable public space throughout all city neighborhoods, poor as well as a rich, however, indicates the need for conventional public open space programs where the incentive zoning strategy does not apply.

Qualitative Results

Although the quantity of privately owned public space produced under the program has been impressive, the qualitative record is disappointing. The study classified the 503 privately owned public spaces by five use categories: destination, neighborhood, hiatus, circulation, and marginal spaces.[12] Based on a site-by-site survey, the study found that more than four out of ten spaces were marginal, that is, they did not serve any public use.

Destination space was defined as high-quality public space that attracts employees, residents, and visitors from outside as well as from the immediate neighborhood.[13] Users socialize, eat, shop, view art, enjoy outdoor music, read, or just relax. The design appeals to a broad audience. Spaces are well-proportioned, lighted, climate-controlled if indoors, aesthetically interesting, and constructed with quality materials. Amenities may include a combination of food service, artwork, regular programs, restrooms, retail frontage, and water features as well as seating, tables, trees, and other plantings (figure 2). The space is well maintained and public use is generally steady.

Neighborhood space is high-quality public space that draws residents and employees on a regular basis from the immediate neighborhood, including the host building and its environs within walking distance. Neighborhood space is used for such activities as socializing, child care, reading, and relaxation. Neighborhood spaces are generally smaller than destination spaces, are strongly linked with the adjacent street and host building, are oriented toward sunlight, and are carefully maintained. Amenities typically include seating, tables, drinking fountains, water features, planting, and trees, but not food service or live entertainment.

Hiatus space is public space that accommodates passersby for a brief stop, but never attracts neighborhood or destination space use. Usually next to the public sidewalk and small in size, such spaces are characterized by design attributes geared to their modest function and include such basic amenities as seating.

Circulation space is public space that materially improves the pedestrian's experience of moving through the city. Its principal purpose is to enable pedestrians to walk more pleasantly and quickly from one point to another. Indoor circulation

Figure 2   Mother and baby in SONY Atrium, New York City. (Photo by R. H. Platt.)

spaces provide weather protection and removal from traffic noise. Circulation space may be uncovered or covered, and sometimes fully enclosed. It is often one link in a multiblock chain of spaces. Size, location, and proportion all support its principal mission. It usually lacks seating and other amenities that invite lingering.

Marginal space is public space that lacks satisfactory levels of design, amenities, or aesthetic appeal, and thus deters public use for any purpose. Such spaces usually have one or more of the following characteristics: barren expanses or strips of concrete or terrazzo, elevations above or below the public sidewalk, inhospitable microclimates characterized by shade or wind, no functional amenities, spiked railings to deter sitting, dead or dying landscaping, poor maintenance, and no measurable public use.

The study classified the 503 spaces as follows:

15 destination spaces (3 percent of the total)
66 neighborhood spaces (13 percent)
104 hiatus spaces (21 percent)
91 circulation spaces (18 percent)
207 marginal spaces (41 percent).[14]

Classifying each of the 503 spaces relied on visual observation and user interviews.[15] Each space was visited more than once at different times of day and night and time of year. A subset of spaces was studied more intensively with additional visits and more rigorous documentation.[16]

Visual observations for each space were documented in text and graphic formats, including written notes, tape recordings, photographs, hand-drawn site plans, and analytical sketches. Observations first focused on how many people were present, what they were doing, where they congregated, which amenities they used, how they entered and left the space, and their demographic characteristics. Next, salient aspects of design and operation were noted, with particular reference to how they supported or discouraged use. Design elements such as size, shape, orientation, location, materials, and amenities were noted in relation to which uses such elements would support. Operational elements involved how the space was maintained, how it was managed vis à vis responsiveness to the public's right to use the space, and whether the space was in apparent compliance with applicable requirements.

User interviews were conducted at every space that had users. Users were asked whether they knew that it was a privately owned public space, why they were there, how often they came, where they had come from, what they were planning to do, and so forth. Users were also invited to make general comments about the space, including what they liked and disliked about it and how it compared with other public spaces.

Calibrating the Law to Improve Design

The record of outdoor privately owned public spaces (plazas, urban plazas, and residential plazas) convincingly demonstrates the power of law to fashion good and bad outcomes, and to be adjusted over time to reflect evaluation of results. The study revealed a chronological fault line in the quality of space created before and after the mid-1970s, when the city significantly amended the original 1961 zoning incentive resolution. To this day, most of the plazas of the 1960s and early 1970s are unusable, unaesthetic, or ill-situated. Of the 167 plazas, 105 (63 percent) are marginal spaces, 37 (22 percent) are hiatus spaces, and none is a neighborhood or destination space. The 1961 Zoning Resolution bears primary responsibility for this result. Although its original goals were to promote access to light and air and public use,[17] the adopted plaza definition favored the former and ignored the latter. The minimal legal standard required only that the space be open and accessible to the public, along with modest design standards. Office and residential developers were allowed to install paving around the base of their buildings, call it a plaza, and collect the 10:1 or 6:1 floor area bonus as a matter of right. The record of these plazas unequivocally demonstrates how they could concurrently satisfy the "letter of the law" yet fall dramatically short of creating usable public places.[18]

Marginal plazas suffer from some or all of a variety of defects. They are environmentally and aesthetically hostile to public use and are typically described as barren, desolate, depressing, and sterile places.[19] They are vacant strips or larger expanses, shaped and located indifferently, surfaced in inexpensive materials such as concrete or terrazzo.[20] Slight elevation changes above or below the adjacent sidewalk often remove them from the life of the street.[21] Their microclimates are unappealing, with surfaces frequently untouched by sunlight and sometimes subject to wind tunnels created by unfortunate juxtapositions of vertical and horizontal planes.[22]

Marginal plazas lack such basic functional amenities as seating, let alone tables, drinking fountains, food service, and programs. Of the 320 commercial and residential buildings with public spaces, the study found that 43 percent have public spaces without any required amenities whatsoever, mostly "as-of-right" plazas and arcades. Ledges that could serve as sittable surfaces often are aggressively detailed with metal spikes and railings or, if unadorned, are too narrow or awkwardly sloped for comfortable sitting.[23] The plazas also lack such aesthetic amenities as landscaping, ornamental water elements, and artwork, which enrich the urban experience. Trees and shrubs are usually scraggly, displayed in unappealing concrete, plastic, or wood planters.[24]

Plazas in front of residential buildings often double as passenger drop-off driveways, entrances to an underground garage, or loading docks. Of the forty "as-of-right" plazas at residential buildings on the Upper East Side, for example, nineteen have driveways.[25] For many years such "private" uses did not invalidate the qualification of that portion of the plaza for a zoning bonus.[26] Plazas are not identified by plaques, signs, or other graphic materials as public spaces, so members of the public cannot know they are entitled to use the space in the unlikely case that they would want to do so.

Many plazas are "a-contextual," randomly situated without due regard for adjacent sidewalks and streets, buildings, and other public spaces. The 1961 Zoning Resolution permitted this result, authorizing the placement of "as-of-right" plazas throughout most commercial and residential high-density districts. Although the goal of light and air in a dense urban setting is laudable, it is not automatically appropriate in every case. The Seagram Building with its celebrated plaza, one of the models for the 1961 Zoning Resolution,[27] operates splendidly on its Park Avenue site between East Fifty-second Street and East Fifty-third Street in part because it is visually enclosed by other buildings. If adjoining sites did not provide a sense of counterpoint and enclosure, then the appeal of Seagram's "tower in a park" would be severely diminished.[28]

That is precisely what happened several blocks to the west, where three towers—1211 Sixth Avenue, 1221 Sixth Avenue, and 1251 Sixth Avenue—all developed as part of the Rockefeller Center complex and designed by the architectural firm Harrison & Abramovitz, planted three plazas in a row on the west side of Sixth Avenue between West Forty-seventh Street and West Fiftieth Street. Ranging in size from 20,000 to 30,000 square feet, these massive spaces provide much light and air, but their juxtaposition also demonstrated that "contiguous plazas which totally obliterate the street wall" and banish retail from the public sidewalk may harm urban vitality.[29]

Zoning amendments in 1975 and 1977 prescribed detailed new design requirements for plazas affecting location, orientation, shape, proportion, elevation, functional and aesthetic amenities, and public identification. The quality of urban and residential plazas accordingly improved dramatically.[30] Developers began to provide spaces that looked more like urban rooms than leftover strips or superfluous expanses. The study found that required seating, planting, trees, lighting, and plaques are located at roughly half of all buildings with public space, principally within the post-1975 urban and residential plazas. Drinking fountains and bicycle parking are found at roughly one of every five buildings.[31] Decorative water features are found in about one-fifth of the sites.[32] Thoughtful design by professionals specializing in public spaces enhances the aesthetic, as well as functional, experience.[33] Sculptures and iconlike structures are commonly installed.[34] Paving and building wall coverings are decorative and varied.[35] Direct sunlight is enhanced through careful site design.[36] New spaces do not create undesirable gaps in the enclosing street wall. As would be expected, post-1975 outdoor spaces are more heavily used than pre-1975 spaces. Of the eighty-nine postamendment urban and residential plazas, the study classified thirty-five as neighborhood spaces, thirty-nine as hiatus spaces, and only six as marginal spaces.[37] This contrasts sharply with 63 percent of pre-1975 plazas deemed to be marginal.

The eighty-eight "as-of-right" arcades have a similarly disappointing record; the study classified sixty-three (72 percent) of them as marginal.[38] The partially or fully covered pedestrian spaces generally fared better. Of the fifteen covered pedestrian spaces, six were classified as destination spaces, three as neighborhood spaces, and none as marginal space. Of the twenty through-block arcades, connections, and gallerias, fourteen were listed as circulation spaces, two as destination spaces, and none as marginal space.

The better quality of these spaces was the result of detailed, case-by-case review by the City Planning Commission subject to legal standards initially more demanding than those for "as-of-right" plazas and arcades. Furthermore, most of these spaces are functionally integrated with their host building, ensuring high levels of usage and accountability by owners to their tenants.

Privatization and Legal Compliance

Although the mid-1970s zoning amendments improved the initial quality of most outdoor spaces and discretionary review generally enhanced the design of indoor spaces, neither arrested the problem of illegal privatization of public spaces. Based on field surveys during 1998 and 1999, roughly one-half of all buildings with public space were found to be noncompliant with legal requirements concerning public access, private use, or provision of amenities.[39] Ironically, the better-designed, post-1975 outdoor spaces and the partially or fully indoor spaces were prominent in this category. Created under more demanding discretionary review, such spaces had more rules to follow and thus more rules to break.

The phenomenon of public space privatization, either intentional or inadvertent, is not surprising. Privately owned public space introduces tension between private and public interests. After receiving the floor area bonus, the owner is left with a space whose public operation may not please the building's occupants. Some owners believe that the use of public space should be limited to the building's office or residential tenants. Others see economic value in shifting physical use of the space to private enterprise. When ownership of a residential space passes to a condominium or cooperative association, the unit owners may not even realize that the original developer received a financial benefit for providing the public space.

The study found that privatization violations typically fall into three categories: denial of public access, annexation for private use, and diminution of required amenities.

Denial of Public Access

A public access violation occurs when legally required access to a space is impaired by management actions. The most typical circumstance has involved spaces behind fences or inside buildings, whose entry gates or doors were locked during hours when the space was legally required to be open.[40] Public access to all or part of a space also has been diminished from time to time by placement of a physical barrier, such as a planter or dumpster, at a strategic entry or corridor location.[41]

Another form of public access violation has occurred when building personnel misinform the user that the space is not public. The presence of guard dogs and security buzzer systems are additional deterrents.[42] Access denials also have been accomplished when spaces are blocked repeatedly, or for extended periods of time, by construction or repair activities.[43] Sometimes, the space is barricaded behind plywood walls, other times, underneath construction scaffolding for prolonged periods of time. The owner continues to profit from the bonus floor area received through the incentive zoning transaction yet is temporarily relieved of the obligation undertaken to obtain the bonus.

Annexation for Private Use

Annexation of public space for private use is a second form of privatization, occurring when an adjacent commercial establishment or other private use spills out without authorization into part of the public space for its private purposes. It is important first to distinguish between legal and illegal commercial uses in public spaces. The Zoning Resolution requires retail frontage along urban plazas and encourages the installation of open air cafés that serve a paying clientele. Commercial activities near or within public spaces can enliven a moribund space. Unauthorized commercial activities, however, may privatize portions of public space, as when an adjacent food establishment practices "café creep," "brasserie bulge," and "trattoria trickle." Movable tables and chairs, waiter service, and, sometimes, planters defining the perimeter of a dining area illegally invade a portion of the public space, and members of the public are prohibited from sitting at the tables unless they purchase food or drink.[44] Restaurants are not the only illegal privatizers; other examples have included a department store and automobile showroom uses.[45] Public spaces have also served as private parking lots for office tenants of the host building.[46]

The most extreme case of annexation occurs when an owner actually builds a permanent structure in the public space itself. When that happens, the space not only is privatized; it simply does not exist. One example involved a residential owner who allowed installation of a permanent structure used by a restaurant in the required plaza area. When the city finally learned about this violation, it devised a plan that permitted the restaurant to remain in exchange for additional plaza space located elsewhere and supplemental amenities not otherwise required.

Diminution of Required Amenities

The third form of privatization, diminution of required amenities, arises when the owner impairs or removes a legally required amenity. In one extreme case, the owner provided no amenities from the beginning, as if the space were an "as-of-right" plaza, even though the owner was in fact required to construct a residential plaza.[47] The space was eventually upgraded with required amenities.[48] In another case, the owner removed all required amenities, degrading an urban plaza to an "as-of-right" plaza.[49] That space is currently the subject of litigation.[50] More commonly, however, violations involve incomplete compliance with requirements for such amenities as seating, tables, drinking fountains, water features, restrooms, and trees.[51]

Movable chairs are also inherently removable. According to William H. Whyte Jr., when movable chairs were first proposed as a required amenity for urban plazas in the mid-1970s, the New York Department of Buildings objected that it would be hard to police such a requirement.[52] In one especially well-documented case, the owner of a hotel removed some of the movable chairs, required by special permit, following a series of thefts from hotel guests that the owner attributed to perpetrators casing the hotel from the chairs! A series of enforcement actions and appeals ensued, eventually upholding the hotel.

Sometimes, existing amenities have been deliberately disabled by the site owner. Use by homeless people has motivated some site owners to install spiked railings and small fences. Other obstructions, such as planters strategically placed on required benches, have been removed following complaints from public space users.[53] Required public restrooms have from time to time been unmarked and locked, rendering them practically unusable.[54] Water features and drinking fountains are often turned off, and management explains they are under repair.[55]

Further, amenities have been installed in ways that impair their usefulness. For instance, the Zoning Resolution requires urban and residential plazas to exhibit plaques that identify the plaza as a public space, list the most important amenities, and specify a contact number for management (figure 3). Required plaques and signs may be slowly obscured by growing vegetation or may never be installed in the first place. Finally, amenities can be impaired or incapacitated by failure of maintenance. The most common example involves trees and plantings, which may die through neglect.

Figure 3   Posted rules for public use of "590 Atrium." (Photo by R. H. Platt.)

Policy Implications: Enforcement and Improvement

This study highlighted the need to enforce legal obligations regarding existing privately owned public spaces. Prior to this study, however, policies concerned with such spaces focused more on revising standards for new sites than on enforcement of rules concerning spaces already created. Changing the political and economic culture that allowed such neglect of existing spaces for the first thirty-five years of the program remains an overarching challenge. Specific policy proposals to address the shortcomings identified by the study are discussed next.

Improvement of Spaces

Policies that encourage or require the improvement of existing public space should be explored. Under current zoning rules, owners seeking permission from the city to close their spaces at night or install an open-air café or kiosk are usually asked to upgrade their space in return. Owners who seek approval for other changes to their public space could be required to make similar improvements. Owners of existing spaces could be offered additional zoning incentives, such as permission to construct additional floor area, in exchange for public space improvements. For some, the idea of using new incentives to fix spaces that have already generated old ones may be disturbing. For others, it may constitute an acceptable trade-off that takes account of the zoning law's underachieving demands from 1961 to 1975.

The city could also compel owners to improve existing public spaces to remedy widespread deficiencies. For example, it might require installation of public space identification plaques in pre-1975 plazas and all arcades, even though they were created under legal standards that had not required plaques. Owners might complain that this step is an ex post facto imposition of a burden to which they never agreed. Of course, government imposes new burdens on existing property rights—or example, installation of fire detector alarms or tougher environmental standards—under circumstances in which existing conditions of property adversely affect public health, safety, morals, and general welfare. A new plaque requirement would promote awareness that a particular space is in fact public. More costly mandates, such as requiring owners to upgrade their pre-1975 plazas and arcades to post-1975 standards, would more likely incur a property owner legal challenge. Such a mandate would be easy to justify as promoting greater public use, but harder to justify as an attempt to secure the raw fundamentals of the original deal.

Enforcement of Regulations

Government programs that rely principally on the private sector to provide public goods and services may be co-opted by private interests if they lack precise documentation of legal obligations, regular monitoring for compliance, and vigorous reaction to violations. In New York, as mentioned previously, more attention was paid to reforming the zoning standards that created them than to ensuring that the public received the benefits it was promised. To plan is human, to enforce, divine!

An effective enforcement regime for privately owned public space requires five elements: reliable documentation, public knowledge, periodic inspections, meaningful remedies, and promotion of public use. In New York City, the first two elements are now in place. The assembly of comprehensive, accurate documentation involved a three-plus-year legal and planning exercise, best characterized as "forensic accounting," to collect, research, and analyze the thousands of documents constituting the legal basis for the 503 spaces created over four decades. The study's complete documentation now resides in a computer database and commercially published book; the database is to be regularly updated to reflect additions and changes to the public space inventory. Through the database and book, public space users have the underlying information necessary to monitor the spaces as supplemental "eyes and ears" to a more formal inspection process and, when necessary, to pursue legal remedies on their own.

The other three enforcement elements remain elusive in the case of New York City. Although periodic *inspections* of spaces to assess owner compliance with applicable legal obligations are essential, the city's Department of Buildings is unlikely to conduct them. Its approach to public space enforcement is complaint-driven. Only then do inspectors visit a space to determine whether a violation is, indeed, occurring. Given the enormous demands placed on the department to ensure that the city's tens of thousands of buildings, elevators, boilers, and other facilities are structurally sound and safe, it is unlikely that a regime of self-initiated public space inspection will ever have a high priority.

Alternatively, the city could contract with a private organization to manage periodic inspections under standards promulgated by the Buildings Department. Although such inspections would be unofficial, they could motivate the Buildings Department to conduct an official inspection. Owners of public spaces could cover the administrative cost of such periodic inspections, as they do with elevator inspections. As with restaurants, the results of public space inspections could be posted on a Web site.

Another approach is to allow owners to engage design professionals approved by the city to certify that a public space complies with applicable requirements. Owner self-certification involving submission of checklist forms prepared under oath is another possibility, despite possible conflicts of interest. Local community boards and civic organizations could organize unofficial inspections of public spaces and report their findings to the Buildings Department and media.[56]

Enforcement requires meaningful remedies once violations of laws are uncovered. Owner complacency will likely change if lawsuits and penalties are credibly threatened. Based on apparent legal violations unearthed by the study's field surveys, the city conducted additional inspections of selected spaces during the summer of 2000 and subsequently brought three civil lawsuits and eight administrative actions against public space owners.

Legal actions by parties other than the city may be helpful as well. Under New York state law, individuals who allege "special damage" resulting from violations of the Zoning Resolution may sue the property owner.[57] (Suits to force the city to enforce the Zoning Resolution are not expressly authorized by New York state law, however.) The Zoning Resolution authorizes but does not require the Buildings Department to enforce the Zoning Resolution's provisions.[58] Private law instruments, including restrictive declarations and easements reiterating some or all of the legal obligations agreed to by public space owners as well as performance bonds, may be employed as part of a "belt and suspenders" approach to public space enforcement. Recording of restrictive declarations and filing of performance bonds are already required in certain circumstances.[59]

Penalties must be sufficiently onerous to convince owners they are not an acceptable cost of doing business. The city has already increased its schedule of fines pursuant to the study's finding that roughly one-half of buildings with public spaces are not in compliance with applicable requirements. In the spirit of "let the punishment fit the crime," future penalties could be adapted to the violation. For example, if an owner privatizes public space, the city might impose a damages penalty equal to the owner's financial earnings from the floor area improperly preempted. Alternatively, the city could temporarily revoke the certificate of occupancy for the bonus space. The city has employed such "literal" zoning enforcement in the past. In a notorious case, a developer was required to remove the top twelve stories from a newly constructed building after the courts ultimately determined that the extra floors violated height rules of the applicable zoning district.[60] The city chose not to accept a cash payment for affordable housing as recompense for its transgression, even though such a solution was urged by parties at the time.[61] The city also has the ability to seek injunctive relief and prison sentence if circumstances warrant.[62]

The final element of effective enforcement is promotion of *public use*. Public use not only indicates that a space is performing well; it also helps a space perform well. As Whyte discovered in his studies of public space, use, even heavy use, almost never deters more use; instead, use begets more use.[63] Members of the public often take a proprietary interest in public space and consider its legally mandated provision to be one of their rights. Public use makes it harder for owners to violate the law and thereby assists the enforcement regime. The city government and civic groups can facilitate use of public space by adopting a stewardship mentality toward its provision and by understanding and publicizing it as one of the city's array of amenities. Is it too much to imagine New York City's privately owned public spaces as a "decentralized Central Park"?

This essay has discussed the results of a study demonstrating how law can significantly affect, for better and worse, the design and use of the built environment. The study examined the effect of New York City's Zoning Resolution on the provision and operation of 503 privately owned public spaces alongside or within commercial and residential skyscrapers. The study found that minimal design standards governing the program's first fourteen years resulted in marginal outdoor spaces and that heightened design standards adopted in 1975 significantly increased quality. The study also found that owners frequently privatized public space in violation of applicable legal requirements and that existing institutional approaches to enforcement failed to arrest such problems. A series of policy changes aimed at improving enforcement of legal obligations were enumerated.

As Holly Whyte taught us, cities are about publicness, seeing and being seen, mixing and avoiding, accidental encounters and planned meetings. The social and practical functions of urban public spaces have lately been downgraded in many places in the face of rampant privatism and decline of civic values, as documented by Robert D. Putnam in his book *Bowling Alone* (2000). Academic conferences now ask the question, Is public space dead? Yet any observer of streets, sidewalks, parks, and plazas in the more vibrant cities understands that such public spaces effectively *are* the city.

Notes

1. Among his many contributions toward making cities more humane, William H. Whyte Jr. helped promote and refine New York City's program of zoning incentives to encourage developers to establish and maintain various forms of public spaces at their own expense. As documented by Jerold Kayden in his book *Privately Owned Public Space* and this essay (using research methods developed by Whyte), the program has yielded more than five hundred public spaces of various types, but with mixed results in terms of public benefit. [Ed.]

2. The City of New York has also used incentive zoning to obtain other types of public benefits, including affordable housing, subway station improvements, and theaters. New York City Zoning Resolution, Sections 23-90 (housing); 76-634 (subway station improvements); 81-00 (for theaters).

3. Implicit in this rationale is that alternative methods for securing small public spaces, such as buying them with money from a city's capital budget, would be less worthwhile or simply unrealistic. Indeed, incentive zoning is credited with being a marvelously creative solution for obtaining public benefits without expenditure of taxpayer dollars, at a time when public sector budgets are increasingly constrained. See Getzels and Jaffe 1988, 1.

4. 483 U.S. 825 (1987).

5. 512 U.S. 374 (1994).

6. Although the United States Supreme Court has never stated that incentive zoning in its purest, voluntary form is subject to the *Nollan-Dolan* line of the Fifth Amendment just compensation clause analysis, it is nonetheless heartening to be able to argue that there is, indeed, an "essential nexus" between the legitimate public interest in reducing congestion and a condition that secures density-ameliorating amenities, as well as a "rough proportionality" between the public space condition and any harmful impact caused by the bonus floor area. See Kayden 1996.

7. Zoning floor area is a defined term in the Zoning Resolution. See New York City Zoning Resolution, Section 12-10. The amount of zoning floor area in an office building is usually less than the amount of "net rentable floor area" as that latter term is used by New York City's real estate industry.

8. The floor area ratio (FAR) is defined as the total zoning floor area on a zoning lot, divided by the area of the zoning lot. Thus, a ten FAR building is ten stories if it completely covers the zoning lot and rises straight up on all sides, is twenty stories if it covers half of the zoning lot and rises straight up, and so forth.

9. The sixteen million square feet of floor area is the equivalent of roughly six Empire State Buildings, the entire office stock of Detroit, 60 percent of Miami's office stock, or more than one-quarter of Boston's office space inventory.

10. The twelfth type, sunken plaza, was never provided by a developer.

11. For this calculation, an average city block is assumed to be two hundred feet by six hundred feet, totaling one hundred twenty thousand square feet.

12. Public space studies employ a variety of lenses to classify public space, and use is one of the most common. See, e.g., Marcus and Francis 1998, 20; and Carr et al. 1992, 79–86.

13. The immediate neighborhood is defined as the host building and other buildings within a three-block radius. See Whyte 1980, 16 (describing an effective market radius for public spaces of three blocks).

14. Each space was placed within one classification only. If the space met the criteria for more than one classification, it was placed in the one that best characterized it. A number of public spaces under construction or alteration at the time the study was completed were not classified.

15. The methodology for classification relied upon the approach of such researchers as William H. Whyte, who proved the value of "firsthand observation" and described how he "watched people to see what they did" (Whyte 1980, 10, 16; see also Jacobs 1985, 8–9, 133–41, describing more generally the value of observation for purposes of urban analysis). Basic aspects of post-occupancy evaluation techniques were followed. See, e.g., Marcus and Francis (1998, 345–56). Judgments about potential, as well as actual, use were made, especially in cases where it was probable that greater public knowledge about the space would result in greater public use.

16. Whyte's study focused on a sample of eighteen public and private spaces. See Whyte 1980, 26–27. This project analyzed all 503 public spaces in the city in the belief that a comprehensive look would provide additional insights, and to fulfill the project's public policy goal of documenting and publicizing the legal requirements attached to every space. Thus, although a core sample of spaces received observational analysis at the level of Whyte's eighteen spaces, other spaces necessarily received less intense scrutiny. For an example of another study that trained its focus on eight public spaces, four in Los Angeles and four in San Francisco, see Banerjee and Loukaitou-Sideris 1992.

17. See Voorhees Walker Smith & Smith (1958, x) (referring to light and air and usable open space).

18. The occasional outdoor space rising above letter-of-the-law performance, either in initial execution or subsequent upgrading, proved to be the exception to the rule. See, for example, 747 Third Avenue (for initial quality) or One Penn Plaza (for voluntary, self-initiated upgrading).

19. As a City Planning Department report summarized in 1975, plazas can be "bleak, forlorn places. Some are hard to get to. Some, sliced up by driveways, are more for cars than for people. Some are forbidding and downright hostile" (New York City Department of City Planning 1975, 5). At least one owner's representative shared that sentiment. In response to a 1986 Department of City Planning mailing about public spaces, with regard to the plaza at 160 East Sixty-fifth Street, he wrote, "I am compelled to advise you that our set-back is merely an enlarged sidewalk with no amenities whatsoever. Further, there are heavily trafficked store and building entrances and exits, and there are a series of steps that could be a trip hazard for people with vision impairment. Therefore, it would be ridiculous to encourage the use of this space" (letter from Robert Hammer, David Frankel Realty, Inc. to Herbert Sturz, Chairman of the City Planning Commission, 28 October 1986).

20. See, for example, the plazas at 95 Wall Street or 950 Third Avenue.

21. See, for example, the plazas at 200 East Thirty-third Street, 178 East Eightieth Street, or 301 East Eighty-seventh Street.

22. See, for example, the plaza at 1114 Sixth Avenue.

23. See, for example, the plazas at 200 East Thirty-third Street or 160 East Sixty-fifth Street.

24. See, for example, the plaza at 885 Second Avenue.

25. See, for example, the plazas at 200 East Sixty-second Street or 220 East Sixty-fifth Street.

26. As a matter of practice, the New York City Department of Buildings began to disqualify that portion of the plaza devoted to such uses for a zoning bonus in the early 1970s.

27. The Voorhees report reproduced a photograph of the Seagram Building and Plaza, with a caption underneath stating, "Open area at ground level permits a higher rise before a setback is required, as well as a bonus in Floor Area Ratio" (Voorhees Walker Smith and Smith 1958, 128).

28. See Kwartler 1989, 201–3 (discussing Seagram Building model for zoning envelope and problems of context).

29. New York City Department of City Planning 1975, 35. William Whyte commented, "The Avenue of the Americas in New York has so many storeless plazas that the few remaining stretches of vulgar streetscape are now downright appealing" (Whyte 1980, 57).

30. See, for example, the urban plaza at 535 Madison Avenue and the residential plaza at 200 East Thirty-second Street.

31. See, for example, the residential plaza at 301 East Ninety-fourth Street.

32. See, for example, the residential plaza at 630 First Avenue and the urban plaza at 40 East Fifty-second Street.

33. Landscape architect Thomas Balsley is the most prolific of the city's public space design specialists, and a plaza he recently redesigned was named by the owner in his honor. (See his essay in this volume.) Other notable designers associated with public spaces in New York City include landscape architects M. Paul Friedman, Lawrence Halprin, Weintraub and di Domenico, Quennell Rothschild Associates, Zion & Breen, David Kenneth Spector, and Abel Bainnson and Associates.

34. See, for example, the plaza at 9 West Fifty-seventh Street and the residential plaza at 300 East Eighty-fifth Street. The role of physical "icons" in city life is interestingly described by Costonis (1989, 47–51); see also Fleming and von Tscharner (1987, 2–3, discussing "the landscape of the mind").

35. See, for example, the residential plaza at 150 East Thirty-fourth Street.

36. See, for example, the residential plaza at 524 East Seventy-second Street.

37. In addition, owners of five "as-of-right" plazas have ameliorated conditions at their spaces—bringing them closer to an urban or residential plaza—as a condition for securing approval for a night-time closing or installation of an open air café. See, for example, the plazas at 810 Seventh Avenue and 1370 Avenue of the Americas.

38. See, for example, the arcades at 180 Water Street and 489 Fifth Avenue.

39. The field surveys were conducted principally by staff for the New York City Privately Owned Public Space Project. Past data from 1900 to 2005, assembled from less systematic field surveys, inspections by the Department of Buildings, and complaints from citizens and community boards, show at least one-third of all public spaces with compliance problems.

40. See, for example, the through block galleria at 135 West Fifty-second Street, the mini-park and public open area at 240 East Twenty-seventh Street, the plaza at 330 East Thirty-ninth Street, and the residential plaza at 200 East Eighty-ninth Street.

41. See, for example, the residential plaza at 182 East Ninety-fifth Street.

42. This situation happened to the author of this article.

43. See, for example, the through-block galleria at 135 West Fifty-second Street, whose frequently locked gates are supplemented from time to time by construction scaffolding blocking access to the locked gates. Years ago, the escalators providing access to the elevated plaza at 55 Water Street would be regularly under repair, although this condition has improved in recent years.

44. See, for example, the plazas at 1700 Broadway and 211 West Fifty-sixth Street.

45. See, for example, the approved permanent passageway atrium at 712 Fifth Avenue or the arcade at 555 West Fifty-seventh Street.

46. See, for example, the arcade at 160 Water Street or the plaza at 299 Park Avenue.

47. See the public space at 340 East Ninety-third Street.

48. See the public space at 340 East Ninety-third Street.

49. See the urban plaza at 40 Broad Street.

50. See complaint in *City of New York v. 40 Broad Delaware, Inc.*, No. 403829/00, Supreme Court of the State of New York, 13 September 2000.

51. See, for example, the residential plaza at 330 East Seventy-fifth Street for failure to provide most amenities, or the removals of water features at the otherwise fine residential plaza at 171 East Eighty-fourth Street and the plaza at 345 East Ninety-third Street.

52. Whyte 1980, 36.

53. In the 1980s, the management of Trump Tower placed a planter on a required marble bench that, following complaints, was removed.

54. See, for example, the covered pedestrian spaces at 60 Wall Street and 805 Third Avenue.

55. See, for example, the water feature at the covered pedestrian space at 805 Third Avenue.

56. The Municipal Art Society of New York City in 2002 arranged for a day of public space inspections by some of its members and has announced plans to make this event an annual occurrence. (These inspections are done by people called "Holly's Rangers" in tribute to Holly Whyte's interest in these spaces.)

57. See *Marcus v. Village of Mamaroneck,* 283 N.Y. 325, 332-3, 28 N.E.2d 856, 859-60 (1940).

58. New York City Zoning Resolution, Section 71-00.

59. See, for example, New York City Zoning Resolution, Section 37-06 (restrictive declarations for nighttime closings); Section 37-04 (k)(4) (performance bonds).

60. *Matter of Parkview Associates v. New York,* 71 N.Y.2d 274, 519 N.E.2d 1372, cert. denied, 488 U.S. 801 (1988).

61. See, e.g., Editorial, "The Best Way to Punish Illegal Building," *New York Times,* p. 30, col. 1, 14 May 1988.

62. New York City Zoning Resolution, Section 11-61.

63. Whyte 1980, 19.

References

Banerjee, T., and A. Loukaitou-Sideris. 1992. *Private production of downtown public open space: Experiences of Los Angeles and San Francisco.* Los Angeles: School of Urban and Regional Planning, University of Southern California.

Carr, S., M. Francis, L. G. Rivlin, and A. M. Stone. 1992. *Public space.* New York: Cambridge University Press.

Costonis, J. J. 1989. *Icons and aliens: Law, aesthetics, and environmental change.* Urbana: University of Illinois Press.

Fleming, R. L., and R. von Tscharner. 1987. *Placemakers: Creating public art that tells you where you are.* New York: Harcourt Brace Jovanovich.

Getzels, J., and M. Jaffe. 1988. *Zoning bonuses in central cities.* PAS Report No. 410 (September): 3–4. Chicago: American Planning Association.

Jacobs, A. B. 1985. *Looking at cities.* Cambridge, MA: Harvard University Press.

Kayden, J. S. 1996. Hunting for quarks: Constitutional takings, property rights, and government regulation. *Washington University Journal of Urban and Contemporary Law,* 50:135–37.

———. 2000. *Privately owned public space: The New York City experience.* New York: Wiley.

Kwartler, M. 1989. Legislating aesthetics: The role of zoning in designing cities. In C. M. Haar and J. S. Kayden, *Zoning and the American dream: Promises still to keep.* Chicago: Planners Press.

Lassar, T. 1982. *Carrots and sticks: New zoning downtown.* Washington, DC: Urban Land Institute.

Marcus, C. C., and C. Francis, eds. 1998. *People places: Design guidelines for urban open space.* 2nd ed. New York: Wiley.

New York City Department of City Planning. 1975. *New life for plazas* (May): 5.

Putnam, R. D. 2000. *Bowling alone.* New York: Simon and Schuster.

Svirsky, P. S. 1970. San Francisco: The downtown development bonus system. In *The new zoning: Legal, administrative, and economic concepts and techniques,* ed. N. Marcus and M. W. Groves, 139–58. New York: Praeger.

Voorhees Walker Smith and Smith. 1958. Zoning New York City: A proposal for a zoning resolution for the City of New York (August).

Whyte, W. H. 1980. *The social life of small urban spaces.* Washington, DC: The Conservation Foundation.

Criteria for a Greener Metropolis

Mary V. Rickel Pelletier

"Sun and Shadow," "Bounce Light," "Water, Wind, Trees, and Light," and "Sun Easements" are poetic chapter headings under which William H. Whyte Jr. outlined the interplay of nature and urban life in his last book, *City: Rediscovering the Center*. To Whyte, the sensory qualities of our natural environment—especially sunlight—are public rights that ought not be carelessly lost to large private development projects. To raise public awareness, Whyte recognized the need for accurate representational models and rendered drawings of proposed buildings to evaluate the effect of new development projects on the local microclimate. Through movies, still photography, and audio recordings, he explored new techniques for documenting elusive social interactions as well as environmental qualities such as sunlight, shadows, noise level, and wind speed. Even as he urged greater architectural innovation and citizen participation in urban design, Whyte sought legal regulations to control the size of new buildings so as to preserve pedestrian experiences of nature throughout the city.

Whyte's insights into the value of sunlight and his interest in the ways zoning regulations shape urban environmental qualities and how that environment may be enhanced through sensitive building and zoning regulations resonate with contemporary efforts to assess the effect of building construction on human health and the natural environment. Such efforts aim to verify what is today referred to as "sustainable" and "high-performance" "green" building. The Leadership in Energy and Environmental Design (LEED™) rating system—established by the U.S. Green Building Council (USGBC)—has set measurable standards for these design and construction practices that exceed conventional building codes and industry standards. Case studies reveal the numerous diverse benefits of high-performance buildings, including substantial energy savings, positive influences on human health, and conservation of natural resources. Increasingly, progressive cities such as Seattle, Austin, San Francisco, Chicago, and Portland (Oregon) are now requiring LEED rating for public buildings such as schools, libraries, and government offices. LEED rating, however, assesses the qualifications of individual buildings rather than the whole urban infrastructure. This essay reflects on Whyte's foresight and notes recent advances in the evolution of "green" criteria for cities.

Sunlit Streets through Building Shape

In his 1988 capstone book *City: Rediscovering the Center,* Whyte explains that a primary goal of New York City's 1916 Zoning Resolution—the nation's first such ordinance—was to preserve sunlight at the street level. Circa 1900, building height was limited by the structural limitations of masonry. Because ongoing improvements in steel frame structures and passenger elevators enable ever-taller buildings, appropriate height limits must now be established legally, according to prevailing cultural values. Whyte recognized that the original premise of zoning was to control building heights so as to ensure daylight at the street level:

> Let us start with a look at the antecedents to incentive zoning. In its earliest form, zoning was for the provision of light. In eighteenth-century Paris the height of buildings was limited to a multiple of the width of the streets—low on narrow streets, higher on wide streets. When New York City instituted zoning in 1916, the same principle was applied. (Whyte 1988, 230)

Yet rather than setting an absolute building height limit (as did cities such as Paris and, at the time, Philadelphia), New York City established a setback approach. These laws allowed building height to increase dramatically provided it tapered in bulk, stepping back away from the street as it rose higher. The setbacks sought to prevent narrow residential streets from becoming dark canyons while allowing taller buildings along wider commercial avenues. Tall buildings, however, filled what had been the open space backyards and alley areas of smaller buildings.

Although the original intent of New York City zoning was to ensure daylight at street level so as to improve the pedestrian experience, over time buildings with a ziggurat or "wedding cake" form became common. The mass of these ziggurat buildings, still seen throughout the city, fill the site to the property line at the street easement, rise straight up to the legal height limit, and then step back and up in successively smaller increments as determined by the zoning formulas. By the late 1940s, several decades of compliance to the city's zoning regulations had resulted in predictably uniform building infill throughout the city. Because few building owners and architects were willing to experiment with architectural form, pedestrians were confined to sidewalks along an increasingly monotonous street grid.

The Lever House, constructed in 1952, altered the formal conventions by providing public space at the street level. A milestone in modern architecture, the Lever House's innovative asymmetrical composition, wrapped in an elegant stainless steel glass curtain wall, was a stunning contrast to the familiar stone facades along Park Avenue (figure 1). Rather than filling the site with the massive building base of a ziggurat, the second floor of the Lever House hovers above street level, allowing public passage beneath the building's mass (Krinsky 1988, 40–45). In addition to ample public open space at the street level, a roof terrace enabled private

Figure 1   Passing under the Lever House, three pedestrians glance up at the daylight.
(Photomontage by M. V. R. Pelletier.)

access to an outdoor terrace that included planting boxes and shuffleboard. Although innovative, the project complied with the city's then current zoning law. Because the Lever House tower occupies only one-quarter of the site, no setbacks were required under the 1916 zoning law. Notably, Charles Luckman, then president of Lever Brothers, was a progressive client willing to embrace the smaller, unprecedented design solution presented by the architect Gordon Bunshaft of Skidmore Owings & Merrill.

In 1958, another innovation in urban design, the Seagram Building and plaza, furthered the appeal of public spaces and spurred subsequent changes to New York City zoning. Architect Mies van der Rohe received a zoning variance that allowed a taller building in exchange for a generous public plaza on Park Avenue. Here again the client's commitment of a key client representative, Phyllis Lambert, the project's director of planning, was essential to achieving this unprecedented relationship between a skyscraper and the city street. In 1961, the City of New York amended its zoning ordinance to provide a case-by-case permitting approach that could respond flexibly to such new design proposals. Building owners and developers were encouraged to include public spaces in their projects through an incentive increase in rentable floor area: ten square feet of office space per one square foot of public plaza provided. This incentive bonus generously allowed building size to increase by 20 percent over the regulated size ratio. Ironically, the Lever House, referenced as an exemplary building by advisors to the 1961 zoning reforms, did not require a zoning variance, as did the Seagram Building. The exceptional architectural form of the Lever House and the resulting open space were dependent on the architect's vision and the property owner's willingness to accept a smaller, unique architectural design.[1]

Within less than a decade, astute observers noticed how incentive bonuses were routinely awarded to mediocre public spaces. Developers profited as floor area increases were granted in exchange for "public" spaces with limited natural attributes such as through-block circulation areas, covered sidewalks, and interior shopping arcades (see the essay by Jerold S. Kayden in this volume). The appeal of a unique pocket of public space within an otherwise consistent urban fabric gave way to an alarming new norm as large, new development projects swept away whole blocks of older, smaller buildings.

In an effort to determine how urban density is rendered desirable through design and how zoning positively shapes the city, Whyte developed empirical study methods that documented the relationship of human activity to specific design features within the urban context. As a consultant to the city government, he demonstrated the need for additional zoning reforms to explicitly require signage, trees, lighting, and seating as legal conditions for public areas given in exchange for bonus incentives. Although the city passed these revisions in 1975, Whyte lamented the lasting environmental damage caused by buildings made bigger by incentive bonuses. As he wrote, "But the larger costs of incentive zoning have been in the loss of the most basic amenities—sun and light. It is a loss that is rarely counted" (Whyte 1988, p. 251).

Accounting for Daylight

With an eye to sunlight, shadow, and the nuances of reflected light, Whyte noted the symbiotic relationship of building size to the successful social atmosphere of sunlit public parks and plazas. Dispelling developers' claims that shadows cast upon the shadows of other buildings were inconsequential to the quality of urban light, Whyte demonstrated how the shadows of each new building cumulatively darken the surrounding cityscape. In response to rampant real estate speculation, he insisted that zoning regulations were needed to protect the public right to sunlight at the street level. He wrote:

> Sun is money. To be able to take away so many units of sunlight is the other side of the coin of being able to put up that many more feet of commercial space. An architect can modify this equation somewhat by the way he configures the building. But the key

factor is bulk. To repeat: big buildings cast big shadows. Bigger buildings cast bigger shadows. And make more money. Unless the city has rigorous guidelines for bulk and sun and light—and the mettle to stick to them—money will win out over sun. (Whyte 1988, 261)

Exploring ways to assign value to sunlight so that it could not be overlooked in development transactions, Whyte considered legal options such as the "transfer of development rights" (used to protect historic landmark properties) and "prior appropriation" (the law governing western water rights), yet he preferred solar easements (Whyte 1988, 277). Easements neither shift increases in building size to other parts of the city nor designate first-come, first-served, priorities. Recorded easements "run with the land" and subsequent owners of the property are bound by them. In recognizing solar easements as essential to the long-term success of solar energy collection systems, Whyte foresaw the basis for emerging solar access laws. Since then, such laws have evolved with respect to solar energy systems. The Database of State Incentives for Renewable Energy (www.dsireusa.org), for instance, currently lists thirty-three state solar access laws and guidelines.

Seeking appropriate legal regulations, Whyte simultaneously envisioned innovations in architectural form that could delightfully increase urban density. Enthusiastically he promoted their potential:

> What is needed is solar zoning. Given a sensible limitation on bulk, we can not only reduce materially the blocking of sunlight but increase the beneficent reflection of it. We can even manipulate and redirect it to places that had no sun before. In the process we may produce some new building shapes, eccentric and effective. (Whyte 1988, 258)

He pointed out how zoning ordinances can be based on calculations of daily and seasonal exposure to sunlight. Rather than approximating pedestrian exposure to sunlight with building setbacks based on street grid geometry, solar zoning specifies setback parameters according to a "solar envelope," calculated with respect to the changing path of the sun throughout the year. Thus, a solar envelope can scientifically shape the seasonal impact of a building's shadow on the local microclimate and thereby specify amounts of sunlight for unique public places within the city.

The solar envelope concept was outlined by Ralph Knowles as an architectural design tool in response to the energy crisis of the late 1970s. At that time, Knowles summarized those interests: "New values emphasizing energy conservation and the effective use of solar resources require a new kind of zoning envelope based on the geometry of the sun's path" (Knowles 1981, 40). Solar envelope calculations can optimize building orientation with respect to solar energy collection systems, and calibrate daylight within interiors. Because of shifts in federal government priorities, however, U.S. national interests in solar energy waned during the 1980s. Today, Knowles emphasizes the human health gains from physical exposure to the daily and seasonal sun cycles. Although there is increasing evidence that the solar envelope can accurately calibrate sunlight to benefit building owners, occupants, and city dwellers, solar envelope calculation guidelines for high-density urban areas have yet to be developed.

Temporal urban environmental qualities, such as how the building's shadow will be cast across city neighborhoods, are rarely depicted in architectural design drawings or addressed within the design process. In 2003, the National Building Museum presented state-of-the-art building projects in an exhibit and book titled *Big and Green: Towards Sustainable Architecture in the Twenty-first Century.* Although the fifty projects presented important achievements in energy efficiency and indoor environmental qualities, none of them provided graphic analysis, such as sun studies, of the effect "big and green" buildings have on their surrounding environments (Gissen 2002). Yet by being bigger, even green buildings impact the ambient environment experienced by urban residents: wind speeds, air quality, rainwater runoff, and a view to the blue of the sky.

In 1988, Whyte noticed that most graphic analysis, such as sun studies, delivered to the New York City Planning Commission for new building permits were incomplete or incorrect. Recognizing that developers, architects, and city planners do not evaluate the effect of building project proposals on the city, Whyte envisioned a financially independent urban research center that could facilitate objective public review of new development projects, so as to improve the urban environment:

> Such a center would be staffed and equipped to apply a wide range of techniques. It would do sun studies using the models and before-and-after methods pioneered by Berkeley; it would do computer mapping of sun and shadow patterns as practiced by several architectural firms. . . . The center would do wind-tunnel testing to determine the drafts a building might induce and the measures that would modify them. It would also study the winds generated by nature and some microclimatic defenses. (Whyte 1988, 269)

Whyte continued, "Its best research tack will be exploring new possibilities, new ways of making microclimates more benign, sun and light more pervasive" (Whyte 1988, 269). Whyte's concept was based in part on the work of the Environmental Simulation Laboratory (ESL) headed by Peter Bosselmann at UC Berkeley (Whyte 1988, 268). Commissioned by the Municipal Arts Society in 1985 to study Times Square, ESL constructed a crude cardboard model for "walk-through" movies, simulating both existing conditions and future development changes allowable by the 1982 zoning laws. After attracting considerable attention, the public presentations prepared by ESL resulted in subsequent zoning laws revisions. The revised zoning regulations aim to preserve and enhance the unique characteristics of Times Square by establishing building height setbacks that widen pedestrian exposure to the sky (Bosselmann 1998,109). Whyte insisted on the value of providing the public with accurate analysis of proposed building projects so as to balance private interests, architectural innovation, and zoning requirements with community rights.

Today's sophisticated computer programs are able to simulate urban microclimates and render three-dimensional "fly-throughs" in highly detailed cityscapes. Yet the most advanced computer imaging programs readily available for Hollywood movie productions, such as *Spider-Man, Gladiator,* and *The Matrix,*[2] are far too costly for community groups, city planners, and even most public universities. Although the impact of big buildings on the urban infrastructure and atmosphere can be carefully studied during the design process, city planning offices rarely have the resources or staff to prepare accurate analysis. Architectural firms working on big development projects strive to present favorable views, not an analysis that might raise concerns about the urban effects of a design proposal. Project teams are paid to successfully present a project to city officials through compelling, often intentionally impressionistic images (Dunlap 2003). New development projects routinely replace the material and formal variations of a cluster of small buildings with the monolithic conditions of bigger buildings. Bigger, often flat building surfaces uniformly reflect and at times intensify microclimate conditions, such as wind tunnels, glare, urban heat islands, and stormwater runoff.

In 1991, ESL prepared an extensive analysis of urban form on Toronto's microclimatic conditions. Bosselmann's research synthesized diverse data from seasonal natural microclimate conditions with the formal characteristics of existing and proposed buildings into a simulation of pedestrian views, wind speed, and thermal conditions. This research informed the Toronto City Plan.[3] Recently, Kevin Settlemyre of the Green Roundtable worked with students to collect and document empirical data on the microclimates of Boston Common and Back Bay to outline beneficial variations in climate-sensitive design (Settlemyre and Thomson 2005). Settlemyre and Thomson used ECOTECT, a computer program developed in Australia by Andrew Marsh. This program synthesizes urban microclimate information with computer programs that analyze interior building daylighting. As Whyte had anticipated, there are ways to analyze the physical repercussions of new building projects on the city's ever-changing infrastructure, yet to date there are no shared sources of such information available to citizens, city staff, designers, or the developers.

Green Roofs

Emerging architectural features such as green roofs demonstrate an impressive range of benefits for urban environments as well as building owners and occupants. Contemporary garden roofs (or green roofs), as refined and tested in Europe, have evolved beyond heavy planter boxes into lightweight roof systems using a granular growing medium formulated to support a thin layer of vegetation. Studies show that vegetation filters air pollutants such as heavy metals, diesel soot, and dust that settle onto rooftops. Absorptive green roofs retain rainwater, thereby reducing the runoff that causes combined sewer overflows in municipal storm drainage systems. By minimizing solar heat gain, green roofs mitigate urban heat island effects, while providing habitat for birds and airborne insects.

In addition, temperatures stabilize beneath the vegetation layers where water-proofing products are protected from the harsh ultraviolet rays of sunlight, which could result in increased product durability. During hours of peak electricity demand, less energy is required for air conditioning below green roofs. Because green roofs absorb external noises, interiors are quieter. Visually delightful, green roofs improve the sensory qualities of interior and exterior building environments.

The U.S. Environmental Protection Agency (EPA) and Lawrence Berkeley National Laboratory, along with nonprofit organizations such as Green Roofs for Healthy Cities and the Earth Pledge Foundation, are promoting interdisciplinary research and discussion among product manufacturers, designers, installers, and city planners. Demonstration projects in various cities are collecting performance data to determine appropriate municipal policies and incentives.

City officials in Portland, Oregon, have initiated an aggressive "Ecoroofs Program" to alleviate combined sewer overflow problems. Drought-tolerant vegetated roofs that are not irrigated (thus the nomenclature "ecoroof" rather than "green") are an approved stormwater management technique under Portland's Stormwater Management Manual of requirements for new construction or redevelopment. Since 1999, about a dozen demonstration ecoroof projects have received municipal grant funding. The Ecoroof Program of Portland Environmental Services provides guided tours of these ecoroofs as well as technical assistance to residential, commercial, and industrial property owners. In addition, the city has provided economic incentives. Ecoroofs meet Portland's public works code requirements for on-site management of stormwater runoff. In specified districts, developers may apply for a floor area ratio) bonus incentive of up to three additional square feet for every one square foot of ecoroof.

In 2001, Chicago City Hall was transformed by a 20,300-square-foot demonstration roof garden designed by William McDonough + Partners (figure 2). Half of the building is used as City Hall; the other half is used for Cook County government offices. Scientists are currently collecting data on the green roof above City Hall to compare with data collected from the conventional tar roof covering the Cook County half of the building (see City of Chicago's website, http://egove .cityofchicago.org, for data). On average, between 10 a.m. and 1:00 p.m. City Hall green roof temperatures are 20 degrees Fahrenheit less than the adjacent conventional roofing surface. On an August day, however, the conventional black tar roof temperature becomes as much as 50 degrees Fahrenheit warmer than the adjacent green roof. At a 2004 conference, "Greening Rooftops for Sustainable Communities," the Chicago Department of Planning and Development reported that more than eighty green roofs covering one million square feet of roofing have now been

Figure 2 The Chicago City Hall green roof. (Photo by M. V. R. Pelletier.)

constructed or are being planned for public and private properties within Chicago's city limits. Chicago's green roofs program began in 1998 when the city was one of five cities selected by the EPA to receive funding and technical assistance for pilot projects intended to mitigate urban heat island effects (the other cities selected for this program were Baton Rouge, Houston, Sacramento, and Salt Lake City). City officials in Chicago now actively promote green roofs and are considering various financial incentives for building owners. Yet green roofs are just one of many available design advances that are environmentally sensitive, energy efficient, and more enjoyable.

Greening the Building Industry

In 2000, the USGBC introduced the Leadership in Energy and Environmental Design (LEED) rating system to provide third-party verification for "green" buildings qualifications.[4] LEED has proven to be an effectively simple means to verify a project achieves an array of advanced design and construction practices. Building owners sensitive to annual energy costs, long-term maintenance expenses, and the increasing threat of liability from "sick building syndrome" are able to require LEED rating, so as to achieve the comprehensive benefits of high-performance design. LEED points are available by meeting the design and construction criteria specified in six categories: sustainable sites, water efficiency, energy and atmosphere, materials and resources, indoor environmental quality, and innovation and design process. Points within each category detail strategies that support environmentally sensitive building conditions. For example, sustainable site points are awarded for the redevelopment of brownfield sites, bicycle storage areas, restoration of natural habitat, on-site stormwater management, and light pollution reductions.

Points are received for quantitative achievements, such as reductions in energy use, as well as for measurable qualitative conditions, such as providing daylight and views to 75 percent of the worker-occupied spaces. LEED is now developing rating systems that refine point specifications with respect to different building projects. For example, the LEED rating system requirements for new construction vary from the requirements specified in the LEED rating system requirements for existing buildings. Written verification of project work related to point specifications is submitted to the USGBC. One of four LEED certification levels—certified, silver, gold, platinum—is achieved according to the number of total points earned.[5]

By setting higher building standards, LEED aims to transform the market culture by integrating the diverse interests of stakeholder concerns within the building industry, including the views of building owners, occupants, architects, real estate agents, environmentalists, industrial hygienists, developers, contractors, manufacturers, and product suppliers LEED even includes astronomers within its diverse coalition of proponents by awarding points to outdoor lighting fixtures that reduce night light pollution.

All too often, building design and construction are driven by market conventions. Dressed in decorative finishes, familiar plans for standardized programs are built for the cost per square footage resale profits without consideration of energy efficiency, healthy interiors, or the surrounding urban environments. The architect and client focus on architectural appearances, while assuming environmental health issues will be resolved by engineering systems. LEED accounts for the less obvious ambient qualities such as energy, indoor air quality, and acoustics that affect human health and productivity. These shared conditions—our "atmospheric commons"—are easily overlooked in the design process.

Scientific Basis for Better Building

High-performance design requires quantifying sensual environmental qualities within the design development process of each new project. Evaluations such as analysis of daylight and energy modeling reveal ambient conditions not depicted in conventional architectural drawings. LEED rating requires computer modeling studies to verify energy efficiency by illustrating the effects of mechanical heating, ventilation, and cooling within simulated spaces.

With such analytical information, designers can measure how distinct parts affect the whole building as a system. The effect of a south-facing window on the heating and cooling system within a specific room can be evaluated with respect to the size of the window; its frame, glass, curtains, and exterior awnings; the orientation of the window; and the proximity of shade trees. Even the effects of paint color can be analyzed and calibrated in relationship to factors such as sunlight and electric lighting so as to optimize energy performance.

Although criticized as a checklist approach to design, LEED rating provides third-party verification of measurable green qualities. A number of LEED points are awarded by achieving higher technical standards than are specified by most state and municipal building codes. Without quantifiable criteria, manufacturers, contractors, architects, and building owners can easily "greenwash" by claiming to be environmentally friendly in promotional presentations without adhering to the values of environmental science during the design and construction process. Most building industry professionals, comfortable with conventional technologies and their own intuitive design approach, resist change by persuading clients that environmentally sensitive design costs more, looks bland, or is not necessary. Yet willing designers have demonstrated an ability to merge high-performance building qualifications into a spectrum of building projects, including surprisingly new architecture. For example, the innovative folded form of the Central Seattle Public Library, by maverick architect Rem Koolhaas, achieved thirty-four LEED points. Whatever the appearance, the City of Seattle requires that all city-funded projects of more than five thousand square feet achieve enough LEED points for the "silver" rating. Projects such as Seattle Central Library demonstrate that higher design and construction standards of LEED are available to buildings of all types and that better building science is not style dependent.

The values of high-performance design and construction are especially significant to schools, where better environments provide measurable benefits to human physiology and thus the learning process. A 1999 study conducted by the Heschong Mahone Group for Pacific Gas and Electric demonstrates improved student performance to daylighting. Test scores for more than twenty-one thousand students from three school districts, located in three different geographic regions, were analyzed in relation to the quality of classroom daylight. Students with ample ambient daylight progressed up to 26 percent faster on reading tests and 20 percent faster on math tests than students in classrooms with inadequate daylight.[6] Although further studies are needed to better define the effect of indoor environmental qualities such as daylighting and air quality on student health, attitude, and test scores, numerous studies document the undesirable effects of substandard school buildings on student test scores and behavior. As a Massachusetts Multi-Agency Task Force reported:

> Poor school building conditions have a negative impact on student performance. Thus the debate is not whether a correlation exists—between a better environment and improved learning—but just *how severe* the correlation is. (Aguto et al. 2000, 12)

Unlike commercial buildings, high-performance green schools can be designed to supplement the K through 12 learning process. For example, the front façade of Clearview Elementary School in Hanover, Pennsylvania, has a large sundial that reflects daily and seasonal changes in the sun's path across the sky. On Commonwealth Avenue in Boston, students at the Media and Technology Charter (MATCH) school monitor solar energy collection data from classroom computers (Gould 2003; Pelletier 2003). (Real-time data on the 20-kilowatt photovoltaic system mounted on the MATCH roof is available online at www.matchschool.org.)

Funding for design and installation of the MATCH school's solar roof and classroom monitoring program was provided by the Massachusetts Technology Collaborative, which selected twenty K–12 public schools to receive supplemental funding for the design and installation of renewable energy systems. To be eligible for as much as $650,000 in additional funding, each school district demonstrated had to abide by high-performance design guidelines from the Massachusetts Collaborative for High Performance Schools Best Practices Manual, referred to as MASS-CHPS. Based on the LEED rating system that MASS-CHPS is similar to, this program refines the criteria with respect to Massachusetts building codes, climate conditions, school activities, and environmental priorities. California, the first state to outline higher standards for schools, has developed extensive online resources available to professionals and the public. In fact, detailed information about better school buildings is available from numerous resources, including the U.S. Department of Energy's Rebuild America program (www.rebuild.gov).

For school districts where new construction or major renovation projects are not pending, educational programs developed for the Green School Project by the Alliance to Save Energy (based in Washington, D.C.) and Youth for Environmental Sanity (based in Santa Cruz, California), teach students to evaluate their own school buildings, and then implement energy conservation strategies.[7] In Berkeley, California, Alice Waters of Chez Panisse Restaurant has worked with the Martin Luther King Jr. Middle School to develop "the Edible Schoolyard" program, which teaches children how to grow and cook their own food. These students host community dinners that reflect the diverse cultural heritage of their neighborhood. Shifting away from "warehousing" education, all these learning programs involve empirical analysis of actual school building environments that often result in physical improvements.

Greening Cities

Increasingly, ordinary citizens are seeking ways to raise awareness about the ways buildings affect nature. The accumulated magnitude of apparently benign norms, such as electric lighting, is causing insidious environmental problems. Research now points to night light pollution as a serious disturbance to the biological clocks of humans as well as the nesting and migration habits of birds and animals.[8] Airborne emissions from midwestern coal-fired power plants used to supply electricity is linked to mercury contamination of freshwater rivers, streams, and lakes throughout the northeastern United States. Dated technology embedded within the design of household goods, neighborhoods, and power plants burden the public with health detriments and unnecessary energy costs. Yet tested improvements that can increase energy efficiency are available.

Changing leadership is emerging at local levels, within schools, universities, neighborhoods, corporations, and cities. In addition to numerous public education programs, government initiatives for LEED rating have been enacted in forty-two cities across the United States, including Seattle, San Diego, Phoenix, Arizona, Princeton (New Jersey), and Chicago, (Herren and Templeton 2006). Seven federal government agencies, including the U.S. Department of State, Air Force, Army, and Navy, now require LEED certification for new construction projects. Governors of nine states, including those of California, Colorado, Michigan, and Maryland, have signed executive orders calling for LEED certification of state-funded buildings. Other states are currently considering similar legislation (Herren and Templeton 2006).[9] These high-performance requirements for civic buildings such as police and fire stations, schools, courthouses, libraries, and government offices will reduce future taxpayer costs for building energy, operation, and maintenance.

The adoption of LEED ratings for public, institutional, and commercial buildings will advance building industry practices and thus benefit building owners and occupants. Yet the accumulated effect of building development has fostered an array of diverse concerns such as open space fragmentation due to sprawl, non-point source pollution within water supplies, loss of family farms, increased rates of childhood obesity and diabetes, increased rates of asthma, night light pollution, and extended driving commutes due to the lack of affordable housing and public transportation. Numerous nongovernmental organizations have prepared practical planning recommendations to mitigate the effect of development on natural environments. For example, the International Dark Sky Association (www.dark

sky.org) offers detailed specifications for urban and rural street lighting. The Trust for Public Land outlines equitable distribution of parks, open space, and greenways within urban areas. The International Council for Local Environmental Initiatives (www.iclei.org) provides guidelines to reduce greenhouse gas emissions.

These guidelines are beginning to influence efforts to rank cities according to quality of life and policies that promote sustainability. The 2005 SustainLane U.S. City Rankings (www.sustainlane.com/cityindex/citypage/ranking/) evaluated twenty-five cities with respect to conditions such as the availability of farmers' markets, recycling, air and tap water quality, the percentage of parks to the total urban land area, and the number of LEED buildings. One of the first rating guides, *The Rating Guide to Environmentally Healthy Metro Areas* (Weinhold 1997), ranks 317 metropolitan areas according to thirteen categories of data.[10] These data, compiled to assist persons with health problems triggered by pollution, identify toxic sites rather than lifestyle amenities such as recreational benefits.

Establishing appropriate policies to balance economic interests, industry conventions, public health, and higher living standards is an evolutionary process. Recent efforts to rank cities can be further synthesized with outside research and EPA data into a comprehensive rating system. As a supplement to zoning and building code requirements, a city rating system will give voters and elected officials a criteria through which to assess shared values and provide incentives for environmentally sensitive development practices.

Designing Civic Delight

As an astute observer and critic of city design, William H. Whyte sought to describe the sources of delightful urban experiences, even in the context of higher levels of density. While advocating government policies to control development, he nevertheless championed architectural innovation and public participation in place of uniform regulatory conventions. Whyte called for comprehensive analysis of urban conditions through empirical observation of site-specific conditions, scientific data, and the use of computer (well ahead of his time). He clearly recognized, however, the urgent need to implement change through experimental design, rather than simply study the future. Optimistically, he noted the possibilities of shaping cities by merging scientific precision with unprecedented design flair, as in this remarkable observation:

> If only we think of them, there are all sorts of things we can do to bend light and reflect it to felicitous effect: a slight canting of a façade to catch the late afternoon sun across it; a panel of white canvas up high to light the dark part of a small park; a spire such as that of the Chrysler Building, which glints at you wherever you are and makes you feel the better for it. We need more follies like this. (Whyte 1988, 275)

Notes

1. According to the October 2002 *Vanity Fair:* "Economically investing in an architectural sensation served the Lever Brothers well. Extensive media coverage and critical acclaim drew visitors to the Lever House where Lever Brothers Company products were on display in the lobby."

2. Paul Debevec created *The Campanile Movie,* a photorealistic aerial view of the University of California–Berkeley campus, from photo-rendering techniques developed for his Ph.D. thesis. Those same virtual cinematography techniques created the "bullet time" shorts in *The Matrix* and other Hollywood movies. See www.debevec.org/Campanile for details.

3. According to this analysis:

> The methods used, in combination, to study the effect of buildings on Toronto's climate included modeling existing and potential development for wind tunnel experiments and mathematical modeling of the human body's thermoregulatory system. An important step in the research was to prepare seasonal maps that noted the exact location where wind and comfort measurements had been taken. The research team analyzed these maps and then changed the model to show potential development under existing planning controls on selected sites. (Bosselmann 1998, 149)

Also see City of Toronto, Department of Planning and Development, City Plan 91, Report 25, June 1991.

4. A rapidly expanding nongovernmental organization, the USGBC provides useful information on high-performance design and construction practices for all types of green building projects. Introduced in 2000, LEED-registered buildings now account for approximately 200 million square feet or 6 percent of U.S. commercial construction (A. Wilson, Environmental Design + Construction, January/February 2005). Recently, LEED has developed specialized point requirements for existing buildings and commercial interiors in addition to application guides for schools, laboratories, and retail and health-care facilities.

LEED was preceded by the Building Research Establishment Environmental Assessment Method, a voluntary rating system that had been available in United Kingdom since 1993, and the Building Environmental Performance Assessment Criteria, which was developed in British Columbia. The LEED rating system, however, evolved substantially through years of volunteer effort by diverse professionals in the building industry as well as representatives from government agencies and nonprofit organizations (Malin 2000). For additional information see www.usgbc.org.

5. For a thorough definition of the current point system, see www.usgbc.org.

6. Heschong Mahone Group 1999 for Pacific Gas and Electric see www.h-m-g.com.

7. See www.ase.org and www.yesworld.org/resources. Other school building learning activities are available through the U.S. Department of Energy's Rebuild America program and the EPA's Energy Smart Schools program.

8. Rating requirements are typically for projects larger than 5,000 square feet that receive public funding.

9. See "Resources—State and Local Governments" at www.usgbc.org.

10. The thirteen categories are air quality; drinking water quality; toxic releases; vehicle travel; aircraft operations; manufacturers; agricultural acreage; military facilities; population density; Superfund sites; toxic transfers; heating and cooling demand. This guide is available only in print.

References

Aguto, J., M. Allamby, M. Chimienti, A. Ma, and M. Peralta. 2000. *Improving student performance in Massachusetts public schools.* Medford, MA: Tufts University, Capstone Project.

Bosselmann, P. 1998. *Representation of places: Reality and realism in city design.* Berkeley: University of California Press.

Dunlap, D. D. (2003). The design image vs. the reality. *New York Times,* 28 September, Sec. 11, 1.

Gissen, D., ed. 2002. *Big and green: Toward sustainable architecture in the twenty-first century.* New York: Princeton Architectural Press.

Gould, K. 2003. Fueled by sunshine. *Boston Globe,* 4 October, F1.

Herren, A., and P. Templeton. 2006. LEED(tm) initiatives in governments and schools. U.S. Green Building Council, 15 February. Available online at www.usgbc.org/FileHandling/show_general_filoe.asp?DocumentID=691.

Knowles, R. 1981. *Sun, rhythm, form.* Cambridge, MA: MIT Press.

Krinsky, C. H. 1988. *Gordon Bunshaft of Skidmore, Owings & Merrill.* Cambridge, MA: MIT Press.

Malin, N. 2000. LEED: A look at the rating system that's changing the way America builds. *Environmental Building News* 9(6)(June): 1, 8–15.

Pelletier, M. V. R. 2003. The 2003 Northeast Green Building Awards. *Northeast Sun* 21(3) (Summer): 3–7, 2021.

Settlemyre, K., and Thomson G. D, 2005. Urban climatology. Paper presented at the NESEA Building Energy conference, March, Boston.

Weinhold, R. S. 1997. *Rating guide to environmentally healthy metro areas.* Durango, CO: Animas Press.

Whyte, W. 1988. *City: Rediscovering the center.* New York: Doubleday.

Websites

Alliance to Save Energy: Green Schools Project
www.ase.org/greenschools/newconstruction.htm
The Green Schools program offers extensive resources, and learning activities.

City of Chicago
www.egov.cityofchicago.org

Collaborative for High Performance Schools
www.chps.net
This website contains information on high-performance green schools, including a Best Practices Manual with details for designers and guidelines for school districts.

Database of State Incentives for Renewable Energy
www.dsireusa.org
Established in 1995, the Database of State Incentives for Renewable Energy is an ongoing project of the Interstate Renewable Energy Council, funded by the U.S. Department of Energy and managed by the North Carolina Solar Center.

The Green Roundtable
www.greenroundtable.org/
Along with other resources, the Green Roundtable website posts the Executive Summary of Boston Mayor Thomas Menino's Green Building Task Force Report.

Green Roofs for Healthy Cities
www.greenroofs.org

Earth Pledge Foundation
www.earthpledge.org
and the green roof resources:
www.greeninggotham.org

Ecoroof Program of Portland Environmental Services
www.portlandonline.com/bes

ECOTECT
www.squ1.com/ecotect/ecotect.html

Edible School Yard Program
www.edibleschoolyard.org

Heschong Mahone Group: Daylighting and Productivity Study
www.h-m-g.com
This frequently cited study shows correlation between daylighting and human productivity.

International Council for Local Environmental Initiatives (ICLEI)
www.iclei.org
This website offers information about what local governments can do about sustainability. ICLEI provides guidelines on reducing greenhouse gas emissions

International Dark-Sky Association
www.darksky.org
Guidelines for street lighting that reduces night light pollution as well as links to research.

Massachusetts Technology Collaborative Green Schools Initiative
www.mtpc.org/renewableenergy/Green_Schools.htm
Detailed, case study information on more than fifteen pilot green school projects (including the MATCH School) for new construction and major renovations are listed on this website.

The Municipal Arts Society
www.mas.org/
The Municipal Art Society is a private, nonprofit membership organization whose mission is to promote a more livable New York City. Since 1893, it has worked to enrich the culture, neighborhoods, and physical design of the city.

New York City Department of City Planning
www.nyc.gov/html/dcp/html/zone/zonehis.html
A brief history of New York City zoning as well as current regulations are posted on this website.

Solar Envelopes: Ralph Knowles, School of Architecture University of Southern California
www-rcf.usc.edu/~rknowles/
The Solar Envelopes website includes papers about the solar envelope and other related topics, with graphics.

SustainLane
www.sustainlane.com/cityindex/citypage/ranking/
In addition to the SustainLane U.S. city rankings, this website offers a spectrum of other resources.

U.S. Green Building Council: LEED (Leadership in Energy and Environmental Design)
www.usgbc.org/LEED
LEED Green Building Rating System(tm) is a program of the U.S. Green Building Council.

Building the Right Shade of Green

Colin M. Cathcart

> *Frodo looked and saw, still at some distance, a hill of many mighty trees, or a city of green towers: which it was he could not tell.* J. R. R. TOLKIEN

Green architecture seems to be a contradiction in terms. By definition, architecture is opposed to nature, because it is through architecture and urban design that we cope with our discomfort here on this earth, keep one another company, and together confront an otherwise inhospitable wilderness. Nevertheless, our design responses are instinctive. Humanity has been successful as a species because architecture lies deep within our nature.

No doubt, there are limits. Over the next century, the builders, designers, and maintainers among us will confront our next great challenge, to tailor our constructed habitat to the now apparent finitude of our earthly context. What might this sustainable [1] habitat look like? Although we seem to retain an aesthetic preference for the landscapes of our evolutionary heritage, we have seldom hesitated to congregate in cities, sometimes very large cities, with populations limited only by contemporary infrastructural technologies. [2]

Green design has often been conflated with wilderness appreciation, grass-roots activism, and "back-to-the-land" austerity, but it is arguable that *sustainable design will be practiced in its most radical form in the very center of our cities.* When we challenge preconceptions of what is natural, what is green, and what is to be done, the ironic but inevitable outcome is that green architecture will become urban architecture.

Sometimes, sustainable design decisions—taken in light of such considerations as practicality, cost, durability, aesthetics—deny familiar images and traditional solutions. Take a simple example: a camel driver's daughter, trying to do her homework at night in the middle of the desert. We might expect to find her using a candle or a kerosene camp light, both of which are dirty, dangerous, and in the long run quite expensive. She would be much better off with a photovoltaic powered fluorescent light, however incongruous that might seem. Many people dislike fluorescent light. Its color temperature is alien to that of the firelight we have known for thousands of years. Despite its energy efficiency, fluorescent light seems unnatural.

Photovoltaics also seem unnatural; they convert sunlight into electrical power, invisibly, without effluent, forever. Photovoltaics are clearly high tech: photons dislodge electrons in a semiconducting film on a glass surface. They were not invented by local action or by thinking small; rather, they were initially developed by big science for the U.S. space program, and big international oil companies like BP and Shell now dominate their production. Photovoltaics are sleek, glassy, and as a power source, strangely motionless. We are much more familiar and comfortable with our traditional sources of energy—muscle power, wind and river currents, wood, coal, and gas—which may all be traced back to the sun over longer and longer periods of renewal. Photovoltaics "short-circuit" this whole chain and convert sunlight directly into electricity wherever it is needed. Utility thus contradicts image. This statement, however, is not to suggest that the other sources of energy don't still have their appropriate applications. A romantic evening should still be spent in candlelight. Yet we may have to reconsider our prejudices, for truly green design may neither look "green" nor feel "natural."

This essay categorizes green design in three "shades": pale green, intense green, and extreme green.[3] It will first offer some criticisms of conventional practices of green architectural design, public park design, and New Urbanist design *(pale green)*. The next section *(intense green)* turns to a discussion of examples where green design is intensified by simultaneously promoting (1) sustainable building design and maintenance, (2) a distilled and heightened experience of nature, and (3) an environmentally responsible urbanism. Then these hypotheses will be tested against the maximum case *(extreme green)*. Could a skyscraper—the most intensive use of land ever invented—be designed as a green building? Finally, which of these shades is the right shade of green for a particular purpose and locality?

Pale Green

To many, the most "green" house is the most apparently "natural" house. Picture a log cabin in winter. There may be snow all round, but it's warm inside: the eaves carry long icicles, reassuring frost forms on the windowpanes, and welcoming smoke curls up from a woodstove within. Perhaps there is a simple outhouse by a stream out back, with firewood stacked beside it, and a rude shelter for a jeep. What could be more sustainable, more green? Well, most readers will realize what's wrong with this picture.[4]

Better might be a house that is consciously designed to be sustainable, like the one shown in figure 1, which my firm designed[5] in 1994 for a mountainside near Woodstock, New York. Many of the cabin's problems are immediately corrected. Instead of leaky log construction, the house's shell is made with airtight structural insulated panels (Cathcart 1996). A curving metal roof protects against winter winds by bending low over the north side, while exposing a full three stories of glass to warming sunlight on the south side. Heat recovery ventilation permits fresh incoming air to be warmed by exhaust air. Gray-water heat recovery from

Figure 1   Pale green: The house at Willow.

showers and sinks preheats domestic hot water and the radiant floors. And in the summertime, the overhanging roof shades the glass, great thermal mass keeps the house cool, and dozens of sashes can be opened to the breezes. Because the owners were willing to push the implications of sustainable design to their formal conclusions, the house does not look much like a traditional country house, much less like a log cabin.

Nevertheless, all four walls of this house are exterior walls, prone to heat loss. I live in an old loft building in lower Manhattan and I need no superinsulation, no "earth sheltering." My dwelling unit is already warmed by my neighbors above, below, and on either side. Run down the list of all the very efficient design systems described in the last paragraph, and a townhouse or an apartment will be more efficient than a single-family house in every case. The efficiencies can be even greater with mixed use, where different functions, staggered over time, can reduce peak loads. This house may be innovative, but the innovations merely serve to mitigate its relative isolation.[6]

Of course, the house is isolated for a very good reason: it offers a direct experience of nature. Its occupants enjoy winter sunrises from sunspaces designed to warm the house passively. A screen porch for family dinners seems tucked into the forest, but affords views of distant mountains. A half-mile driveway and power lines had to be carved through the forest to get here, however, and a quarter-mile curtain trench was dug to protect the house from runoff. Several acres of forest have been cleared for the house, for its large septic field, and to open the view of the mountains. An old swamp at the bottom of the hill was excavated to provide a more picturesque pond. Human contact with nature is rarely without environmental impact: people cannot help "improving" their land to suit their tastes.

The owners are refugees from lower Manhattan, and they often drive considerable distances to enjoy the cosmopolitan atmosphere of Woodstock and New York City. Both adults will use their cars many times a day, sometimes just to get coffee, and nondrivers (children, visitors, the elderly) must be chauffeured to all their activities outside the house. Still, this house is a good example of "green design" as it is conventionally practiced. There are many other good examples on the website of the U.S. Green Buildings Council[7] located way out in the country or locked in the midst of sprawl, hopelessly compromised by their beautiful locations. These buildings exemplify *pale green* design.

By contrast, New Yorkers use their subways at a fraction of the environmental impact, with convenient access to all the services of the metropolis. The young and the elderly can get around on their own. Environmental impacts are moderated dramatically—in terms of energy efficiency, habitat conservation, walking and transit use—simply by living in a city (Owen 2004). In postwar America, the most acute observers have shown the present limitations on urban density as having to do with the balance between personal privacy and passively regulated public space (Jacobs 1961; Whyte 1980). New Urbanist and smart growth advocates (Calthorpe 1993; Yaro and Hiss 1996; Duany, Plater-Zyberk, and Speck 2000) contrast sprawl and auto dependence with the attractions of traditionally compact U.S. towns and cities. Gasoline consumption, to take just one example, appears to decline exponentially with increased urban density (Newman and Kenworthy 1989). But the urban lifestyle isn't consciously green. The casual experience of nature is denied to most city dwellers. Dense urban areas will often have minimal daylight. Without green building design and without access to natural experience, existing cities must be considered "pale green" too.

City residents can get a good dose of nature in public parks. The pioneers of modern park design—such as Joseph Paxton, Andrew Jackson Downing, and Fredrick Law Olmsted—intended them to be restorative, palliative, a peaceful contrast to the jangle of industrial city life. Parks, though, are by no means "natural." The best parks exaggerate nature. They are almost "hypernatural": artful concentrations of plantings and landforms, designed professionally, maintained with great care and expense, and used intensively. Many city parks are happily crowded every day. Whatever the intentions of their designers, these parks inevitably trigger memories of picturesque landscapes, of countryside once glimpsed from a car window, of nature witnessed on television, or stories told of pioneering ancestors. They fulfill a need as essential as our need to congregate in cities. Parks represent an urbanized intensification of nature. If located in the city, an actual prairie, an actual forest or tundra or jungle would be far too fragile if it was small enough to fit and far too boring if made large enough to be self-sustaining. A park's esthetic contrast between living nature and its urban context holds drama and meaning. A nice green park, though, if it remains locked within an environmentally noxious industrial city, can represent only a part of the solution. So by themselves, parks, too, are only "pale green."

All three aspects—sustainable building design, natural experience, and urban density— must be pursued together. So rather than extending our sprawling habitats, we should instead intensify the use of lands we have already claimed, reducing auto dependence, building community and public health with opportunities for outdoor exercise and intensified contact with civilized nature. Not only is the urban experience most entertaining when it is most intensive (Whyte 1980), but true sustainability will be most easily achieved in this setting.

Intense Green

This section reviews four specific examples of green architecture at urban densities, two in Europe and two in New York, where all three of the following aspects are in evidence:

1. *Sustainable architecture* strives for a functioning building stock that does not impose a net environmental cost on future generations or elsewhere in the region or on the globe, minimizing ecological footprint effects in both space and time. Sustainable buildings encourage physical health through their design and promote energy efficiency, water conservation, renewable energy use, material recycling, and indoor environmental quality and comfort by taking passive and active advantage of bioclimatic resources, reducing effluents, and allowing buildings to participate in "natural" cycles. Sustainable buildings, however, are not good design if they do not also contribute to the hypernatural experience.

2. *Hypernatural experience* is the intensive, designed, and often quite urban experience of flora, fauna, and the outdoors; recreation grounds testing our own bodies against time, space, and the bodies of others; and parks and compacted landscapes, including the contemplative value of distant views of sky, celestial objects, changes in the weather, water currents and sounds, the rain, and the hydrological cycle. A modest window box is hypernatural, as are community gardens, indoor plants, botanical gardens, and even zoos. This aspect of green is clearly palliative and artificial, but it cannot be faked. Human nature being what it is, hypernatural experience is essential to render green design attractive, ultimately producing ecological urbanism.

3. *Ecological urbanism* is an environmentally responsible pattern of human settlement, which manages territorial resources, traffic flows, and infrastructural sys-

Figure 2 Intense green: Isselstein townhouses (built 2002).

tems in a clear and visible relation to natural systems. One could say that it is humanizing nature; more accurately, it could be termed the rendering of a natural ecology as human artifice (Ingersol 1996). Ecological urbanism is the sustainable habitat of a human-centered ecology. This notion may be simply the well-designed product of the prior two green aspects *(a* x *b* = *c*), but a certain vital density is also required.

The synthesis of these concepts may be exemplified by four projects. The first is a housing project in Isselstein, near Utrect in the Netherlands (figure 2).[8] The Dutch have designed every inch of their little country, claiming new land from the sea and creating dense, sustainable landscapes. "What is nature and what is artificial?" asks landscape architect Dirk Sigmons, "You can't say. The landscape is an abstraction. . . . It is a form of degenerated nature, but at the same time it is a beautiful landscape" (Winner 2002, 48). This paradox typifies *intensive green* design.

Each townhouse has a solarium clad with photovoltaic glass panels, endowing a sun-space with sun power. Each dwelling has a little garden, and also a little bicycle shed. There are more bicycle parking spots in the Netherlands than car parking spots, in part because everything you need is designed to be within biking range. The Dutch design new neighborhoods in terms of transit, mixed use, and open space preservation (Beatley 2000; see also Beatley's essay in this volume). Nothing is left to chance or to change, and with such a static human ecology, a certain cultural sterility results. The architecture is sustainable, the landscape is hypernatural (as is all the Netherlands), but ecological urbanism should invite more of the kinds of choice, conflict, open-ended change, and quick reallocation of resources found in nature. Any given street corner in New York City is far riskier, far more like a natural ecology, with much more to gain or to lose, than this deliberately predictable new town.

In Hamburg, Germany, the original street façade of this office block had begun leaking. The owners were going to replace it, a very disruptive process. We said, don't replace it; just hang a veil of photovoltaic glass in front of it.[9] Not only is the leakage problem solved, but solar electricity is generated along with passive heating and ventilation, and natural light levels are maintained inside the building. Daylight is law in Germany. Office workers may only work where there are certain levels of natural light. Between the building's "skins," the old and the new, two spaces were designed: a second floor winter garden and a street-level sidewalk café. Our buildings and cities must undergo just this sort of green retrofit, where building systems are made more efficient and sustainable and at the same time better attuned to their urban context.

The third example is a new photovoltaic glass canopy for the Stillwell Avenue subway terminal[10] on Coney Island in New York City (figure 3). The design is both old and new, with photovoltaic glass panels mounted into a neo-Victorian steel structure. The expansive daylighted space celebrates mass transit and marks the island's economic revival. With all the honky-tonky thrills Coney Island now stands for, let's not forget what started it all: a broad ocean beach within a subway token's reach of every city family. It's a hypernatural experience on a budget.

As a fourth example, we are designing an environmental learning center in a new waterfront park at Stuyvesant Cove on Manhattan's East River,[11] where we hope to blur the boundaries between building, water, city, and landscape (figure 4). The building is to be passive solar, well insulated, and substantially daylighted; powered by sun, water, and wind. A screen of deciduous vines surrounds the second floor, controlling solar gain according to the seasons. The roof sawtooths are made with photovoltaics on the south-facing slopes; the north faces are ventilating clerestory windows. By generating all the power it needs and treating all the effluent it generates, the building will have net zero environmental impact.

The design also endows an old industrial landfill with rare urban views of sky and water. It creates a place for an artificial wetland to be used as a recreational water park. An atrium greenhouse allows the park to pass indoors, and a café underneath the building allows views of the river and the new landscape.

Because the building is just a little bigger than a New York brownstone and a little smaller than a tenement apartment building, it provides a familiar scale for

Figure 3 (*Top*) Exterior and (*bottom*) interior of photovoltaic glass enclosure above Stillwell Avenue subway station, Coney Island, New York.

Figure 4   Intense green: Stuyvesant Cove Environmental Learning Center (project).

urban ecological education. On the second floor of the building is its major display, an "eco-neighborhood" where all the environmental impacts of everyday urban life will be demonstrated. Outside, the waterfront esplanade connects all the way from downtown to Manhattan's Upper East Side, a new thoroughfare for health-conscious commuters, joggers, and roller-bladers. Net-zero sustainability, hyper-natural wetland landscaping, and the explicit demonstration of urban ecology qualify this design as "intense green."

All four examples are located in the middle range of the scale of urban density. One of the measures of density is floor area ratio (FAR), the ratio of a building's total floor area to that of the plot of land upon which it sits. For example, if a five-story loft building is built right up to the lot lines on all sides, it would have an FAR of 5. Brownstone neighborhoods have about a 2.0 FAR. Suburban neighborhoods are often built below 0.1 FAR. The house near Woodstock, New York, at the rural extreme, has an FAR of less than 0.01.

The Empire State Building has a floor area thirty-two times its plot area (32.0 FAR). Maximum urban densities since the 1960s are often set at 15.0 FAR, which is still very high. Could extremely high densities such as these ever be considered "green"?

Extreme Green

The curators of the National Building Museum in Washington, D.C., invited our firm to speculate on the environmental design of very large buildings for an exhibition they were planning entitled "Big and Green."[12] Given the timing of this assignment, we could not help but be influenced by the design of the World Trade Center, whose destruction our entire staff had witnessed. Despite the ensuing skepticism concerning tall buildings in terms of safety, desirability, and symbolism,[13] we remembered Louis Sullivan's artistic appraisal of the skyscraper:

> It is not my purpose to discuss the social conditions; I accept them as a fact, and say at once that the design of the tall office building must be recognized and confronted at the outset as a problem to be solved—a vital problem, pressing for a true solution. (Sullivan 1896)

We decided to investigate the trajectory of environmental technologies as applied to tall buildings. Pushing to the extreme, what might a green skyscraper look like in the year 2020?

The invitation from the National Building Museum stemmed from our role as solar design consultants for the Condé Nast tower at Four Times Square.[14] The developer, the Durst Organization, was determined that this building would be the first green skyscraper in the country. The design, by Fox and Fowle Architects, deployed an impressive array of sustainable design techniques: recycled construction materials, efficient and finely tuned mechanical systems, water conserving bathroom fixtures, fuel cells providing power for energy-efficient lighting at night, high fresh air rates for excellent indoor air quality; it would take a book to describe all these strategies (Lippe 1998).

Although photovoltaics on the roof had been rejected as too costly, the design envisioned expensive spandrel panels[15] for the glass exterior wall. Working with the construction contractor, Tishman Construction Corp., and the New York State Energy Research and Development Authority, we demonstrated that substituting photovoltaic panels for these spandrel panels could be cost effective.

This building's cylindrical electronic billboard participates in the gaudy light show of Times Square, and with an FAR above 30, it is certainly a high-density urban structure. In terms of the hypernatural and ecological urban aspects just defined, however, this building earns little credit. Because it is a large, monofunctional office block, its mechanical systems can be finely tuned for its use, but there are external inefficiencies; large single-use buildings contribute to overloaded transit systems (rush hour congestion) and utility services (peak cooling demands) during weekdays, followed by extended periods of underuse. Even if an office building is fully rented, it is nearly empty all night and on weekends (that is to say, most of the time). Mixed-use buildings and neighborhoods produce a far better, greener use of urban infrastructure.

This building's experience is not hypernatural either; many people work inside without the daylight and views that are every German worker's legal right. Because it occupies a small lot, hemmed in by Broadway, Forty-second Street, and Forty-third Street, the building makes no contribution to urban open space. Transit access is very good here, and so the density permitted under the zoning is very high.

Farther east on Forty-second Street, an even better office tower is now under construction at One Bryant Park (figure 5) (see figure 2 in Eugenie Birch's essay in this volume). After ten years of experience, green architecture is now practiced on an expert level, and recycled building materials, renewable and cogenerated energy, peak cooling load reductions, rain and gray-water harvesting, and so forth—all the good green strategies—are represented in this design. The designers, Cook and Fox, have not neglected natural experience. A winter garden fills the lobby, and there are wonderful views of Bryant Park to the south. With careful accounting, this new building can aspire to the highest USGBC rating. With both these towers, however, their small properties precluded the creation of significant open space at street level. That is unfortunate in terms of both policy and design.

Why can't we build parks with our towers? The early modernists made exactly this argument: dense cities should be built vertically, with the ground cleared for greenery, parkland, and public amenities (Le Corbusier 1925). At the other end of Forty-second Street, Le Corbusier left a generous plaza for the United Nations' headquarters, as did Mies van der Rohe at the Seagram Plaza on Park Avenue. Despite New York's narrow street grid of 1811 (Plunz 1990)—which assumed brownstone development, not high-rises—both architects designed high-rises with magnificent open spaces. Indeed, Le Corbusier's famous criticism of lower Manhattan (Koolhaas 1979) was not that it was too dense, but that its skyscrapers were too small and too close together. Mies's and Le Corbusier's vision was enshrined in the plaza bonus of New York's 1961 zoning law, but the buildings and projects that resulted—often combining several blocks into "superblocks"—were usually monofunctional housing or office developments (see the essay by Jerold S. Kayden, this volume). And like any other monofunctional project, for more than half the time they were deserted. Observing this desolation, Jane Jacobs's devastating critique of modern high-rise plazas and "tower-in-the-park" developments led her to recommend that cities have many short blocks instead (Jacobs 1961). Reinforced by the New Urbanism's preference for traditional street patterns, the small urban block has become dogma. Misguided by these misaligned principles, we do not build parks with our towers because we have decided that streets are more important. One area of popular consensus for the redevelopment of the World Trade Center site (15 FAR), for example, is that the pre-1970 grid of streets should be reinstated across the site.[16]

Figure 5
Extreme green:
One Bryant Park
(projected 2007).

That would be fine if we were going to build brownstones there. Some of the conflicts between high density and small blocks are illustrated by my proposal for loft development west of New York's garment district (Yaro and Hiss 1996, 122ff). We postulated that the transport infrastructure west of the garment center could easily support higher-density development, if that density were to support an intensive mixed-use district. You can convert a loft building to just about anything. Loft apartments, hotels, offices, and industry can all be accommodated in the same loft building type.

The problem, at this density, is the streets. There is no relief here from chock-a-block bulk. Under conditions of increased urban density, streets become increasingly dark. Even on a sunny day, approaching the garment center from the west is like going into a basement; you proceed from sunlight into darkness and from open streets to anarchic congestion. At this FAR, superblocks provide superior design possibilities—just as Le Corbusier and the early modern architects always said—for good-quality public space, more natural light and greenery, and better traffic management.

There are positive examples in most cities. In New York, Bryant Park is on a superblock, and Rockefeller Center is a megastructure. Park bonuses on superblocks can do for urban development what cluster zoning[17] does for suburban development: preserve significant areas of open space for the public to enjoy, while still satisfying Jacobs's concerns about scale.

Superblocks also allow urban designers better access to on-site natural resources. Nature does not stop at the city limits. Sunlight falls with more than a thousand watts of power on every square meter on earth, cities included. Wind is another powerful resource, especially at high elevations. And just because we intend to use these resources intensively does not make that use any less "natural" (or "hypernatural"?). Other species also use resources intensively. The barossa termite of South Africa, for example, occupies anthills of far greater density than any human city and designs those anthills for completely passive solar heating and earth cooling.[18] Central locations are often well served by transit, and where human cities approach anthill densities it must be considered the equivalent of a natural resource, to be managed, harvested, and improved with care. So the three aspects of intensive green still apply at this extreme end of the scale: sustainable architecture, ecological urbanism, and hypernatural experience.

Our tall building design for the National Building Museum did not apply to any specific site; figure 6 shows a generic large city, with both Chicago's Sears and Malaysia's Petronas Towers on the skyline. Instead, the design focused on the issue of great height: one hundred and fifty floors. The building must be as safe, or safer, than a conventional building, so instead of two fire stairs, this building would have seven. Our engineers, Ove Arup and Partners, modeled a ductile, triangulated structure to withstand the stresses of a major disaster, whether natural or manmade.

By 2020, many trends in technology will achieve complete sustainability. New glass coatings and suspended films—photovoltaics are just one of many—are steadily rising in efficiency and dropping in cost. They suggest new design paradigms. Conventional sustainable buildings are "fat," with as little skin area as possible in relation to cubic volume, so as to minimize heat loss in the winter and solar gain in the summer. If progress on advanced coatings continues, then this principle will be reversed. Buildings will be "skinny" so that selective cladding systems can

Figure 6
Extreme green:
The 2020 Tower
(project).

provide daylight, ventilation, and solar energy to every square foot of interior space. Natural lighting, insulating films, and superefficient mechanical systems will in turn reduce energy consumption. With photovoltaic coatings on vision glass and spandrels and a "tiara" of wind turbines crowning the top (shielded by photovoltaic louvers to eliminate strobe effect), the building will create as much power as it consumes. So, even in the heart of the big city, energy demand may converge with renewable on-site resources, and this big building would operate without any energy impacts.

To purify wastewater, there would be a Living Machine® [19] system at every thirty stories, greening the edge of each setback terrace and public upper floor with pleasing foliage. The plant and bacterial life in these Living Machines would treat, purify, and recycle all water-borne effluents from toilets, sinks, and baths in the building through a series of hydroponic biological processes. These processes also support lush tropical plant life. Water can be used, treated, and reused within the building so that the only losses would be from evapotranspiration and the occasional spill. Now *that's* extreme green!

On sunny summer days, this building would have to sell the extra power it generates to its neighbors. At night, it would have to buy it back. It would therefore depend on a distributed local network of energy production and delivery. On-site water supply will not always be balanced with on-site water demand, either. In other words, the building design implies a symbiotic relationship with its bioregion.

It would not be just an office building. Most floors measure only fifty feet from window wall to window wall,[20] partly for sustainability but also for multiuse flexibility. As the market for citywide space demands, the building will easily accommodate an ever-fluctuating mix of commercial and residential uses. It is an open urban ecology.

The elevator systems also permit great flexibility, with express stops (like the New York subway) every thirty stories. The "sky-lobbies" at these stops could be public places, rimmed with the luxuriant plant growth of Living Machines, multistoried, ramped, with access to generous exterior terraces. On 30, there could be a major hotel lobby; on 60, a multiplex cinema; on 90, a health club; on 120, a winter garden. Mixed use would make this megastructure lively, public, and efficient to run around the clock.

The mixed use is not only vertical; it is also horizontal because the building's footprint is far larger than a city block. It would be knitted into the fabric of a typical urban grid, extending the streets with a system of commercial skylit pedestrian galleries for ground-floor retailing and multiple points of access, and major subway station hall.

Why go 150 stories? Most downtowns are already heavily shaded, so this additional impact would be marginal. Perhaps the most important potential benefit is that tall buildings can save space for parks large enough to permit sunlight to again penetrate to street level. This fourteen million square foot complex on a superblock of nine hundred thousand square feet represents an FAR of 15.5, which is very high density, as befits a big city center. Yet because all this space is distributed over 150 skinny stories, the building leaves public space for courtyards and gallerias, and best of all, for a public park of some seven acres, with neighborhood parking and loading areas located underground. This precious green space in the very center of the city would go a long way to producing a hypernatural experience both outside and inside the building. Because the building is "skinny," no one can get more than twenty-five feet from natural light, operable windows, and spectacular views. Unlike in many office towers, its occupants would know when it rains or when a storm front grows on the western horizon. Doesn't everyone want an outside office or a bedroom with a view?

So, although it's a bit of a stretch, extreme green can be done, even in the busy centers of very large cities. This building would have zero off-site energy and effluent impacts. And a new park could come with it.

Environmental values must inform our technological choices, whether in thoughtful design decisions on specific projects or in the mass-scale decisions we make together, as citizens and consumers. Not all technologies are created equal. As we have seen, technology strains to resolve ecological concerns at both ends of the density spectrum; a house in the country and a skyscraper downtown must both go to heroic lengths to be green. Stand-alone houses must be heavily insulated and sealed against weather extremes, must achieve energy efficiency despite low occupancy levels, and the automobile dependency they encourage must be somehow offset or mitigated. At very urban densities the design challenges are different but equally pressing; here designers must consider abandoning traditional street grids, forgo the conventional wisdom of "fat" energy conservers in favor of "skinny" energy producers, and invent elaborate means to introduce greenery and daylight.

Between these extremes, the "intense green" projects—urban, but not downtown—represent the most reasonable (and, presently, the cheapest) focus for sustainable design. Between 0.5 to 3.0 FAR, buildings can be optimized for daylight, solar power, energy conservation, rainwater collection, and waste collection and treatment. At these densities, simple sustainable strategies may resolve multiple issues, including the provision of hypernatural experience: Ijsselstein's solaria, Hamburg's winter garden, Coney Island's subway to the beach, and Stuyvesant Cove's wetland water park. At these densities, people can choose between multiple transit modes and housing types, and urban habitats can be harmonized with renewable resources. The examples given in this essay are admittedly anecdotal—just a half a dozen projects—but it would appear that green design should be neither pale nor extreme. *Intense green* is the Goldilocks solution—it's just right—the right shade of green.

We humans are not like termites, beavers, or bees. Our energy and infrastructural technologies are still evolving. As a result, we have not yet settled on the design or the distribution of our "native" habitat, and it is difficult to forecast any final point of equilibrium. Over the next century, the edges of our cities may densify; the centers may diversify. Some of us will still want to live in isolation, despite the environmental damage we tend to do there, and some will want to crowd together in the very centers of our cities. Whatever a sustainable human habitat will look like, it is likely to be ecologically urban, architecturally sustainable, and intensively natural. Our skylines will then seem as "natural" as that of Tolkein's mythical City of Lothlorien, which at first sight so confused Frodo: Are those naturalized buildings, or humanized trees? Ultimately, whatever defines our cities will define our species. We will finally recognize our cities as our wilderness, our habitat, our very nature.

Notes

1. Meeting the needs of the present generation without compromising the ability of future generations to meet their needs (WCED 1987). Donald Watson (Watson 2001) discusses the origins of the notion of sustainable development and its official and evolving definitions. Clare Palmer (Palmer 2003, 18) summarizes objections. Human societies have confronted finitude before—the Greek polis, for example (Kitto 1951, 64–79), and the Easter Islanders (Ponting 1991, 1–7)—and this current confrontation is real: we do not have a second or third biosphere to absorb our environmental impacts, as would be necessary if the whole world adopted the current "first world" footprint (Rees and Wackemagel 1991).

2. Although our species evolved in a natural setting—the savannahs and forests of Africa—humans have always tended to live together in ever-larger cities. Aristotle believed that true humanity was only possible in cities (Aristotle's *Politics* 1253), and now half of all Greeks live in Athens. Not everyone will want to live in cities (many have no choice about where they live), but that is now the norm for the majority of the earth's human population (Davis 1966). This "natural" tendency has always been limited: first there were food and transportation limitations, then sewage treatment and public health issues, and now public order and perceived congestion. There have been other hesitations, too. Europe in the Dark Ages was antiurban because cities were considered not only vulnerable to barbaric invasion, but also congenitally sinful (Augustine, *City of God*, Book XXI, 1958).

3. Although there are many architects whose work could as easily illustrate this chapter, all these examples are taken from the work of my design firm, Kiss + Cathcart, Architects (www.kisscathcart.com), if only because I am most familiar with both its strengths and weaknesses. My thanks to Gaby Brainard, Ryan Byrnes, Tony Daniels, Brooks Dunn, Luis Estrada, Kimbro Frutiger, Robert Garneau, and Claire Miflin of Kiss + Cathcart, Architects, and their consultants Arup, Drew Gillette, Atelier Ten, Goldstein Associates, and Judith Heinz. Most of all, thanks to Gregory Kiss, who served as principal-in-charge for many of the projects shown and described here.

4. Logs look natural, but in a cold climate, they are neither airtight nor sufficiently insulating. Icicles indicate insufficient roof insulation and ice-damming. Windowpane frost indicates condensation on too-cold interior surfaces. Wood stoves pollute. The outhouse will poison everyone downstream, and the jeep is exempted from current fuel consumption regulations.

5. Engineering: Goldstein Associates, structural; Drew Gillette, mechanical.

6. Relative data on the environmental effects of various consumer behaviors, transportation, housing, and design choices can be found in Brower and Leon 1999.

7. U.S. Green Buildings Council (USGBC) certifies building designs as being "green" through the use of a checklist called LEED(tm) (Leadership in Energy and Environmental Design). Each green design strategy is awarded credits, and the credit totals allow a building design to be "certified," "silver," "gold," or "platinum." One difficulty with LEED is that the relative economic costs and environmental benefits of the various green strategies are not further weighted. A list of USGBC certified projects may be found at www.usgbc.org. More extensive case studies of these and other green designs may be found at www.eren.doe.gov/buildings/highperformance/projects_list.html.

8. Isselstein, Netherlands, completed 2002. Hans Van Swieten, Associate Architect.

9. HEW, Hamburg (project) 1998. Hamburgische Electricitäts-Werke AG, Associate Architects: Somer and Partner, Berlin; Dr. Lutz Weisser. Engineering: Ove Arup & Partners, New York and Berlin.

10. MTA/New York City Transit Authority, Owner. Jacobs Engineering Group, civil, structural, and electrical engineering. Domingo Gonzalez Associates, Lighting Design.

11. Community Environmental Center Inc., client, Terrasolar, Inc, donor. Ove Arup & Partners, Consulting Engineers USA, and Judith Heintz Landscape Architecture.

12. www.nbm.org/Exhibits/upcoming.html. National Building Museum: Big and Green: Toward Sustainable Architecture in the Twenty-first Century, 17 January–22 June 2003, Washington, D.C.; David Gissen, curator. Fall 2003: Museum of the City of New York, New York. Winter 2003/2004: Yale Art Gallery, New Haven, Conn. See also Gissen 2002.

13. For example, "Reaching the sky, and finding a limit; Tall buildings face new doubt as symbols of vulnerability" by D. Dunlap and J. Iovine, *New York Times, 19 September 2001,* A20; however, "The future of up" by W. Rybczynski, *New York Times, 9 December 2001, Sec. 14, 1,* and "Skyscrapers are here to stay, says panel of experts," *New York Times,* 12 November 2001, E1.

14. Fox and Fowle Associates, Architects; Cosentini Associates, LLP, Engineers.

15. Spandrel panels are opaque glass panels between the top of a window and the sill of the next window above, sometimes giving the appearance of a sheer all-glass skin.

16. For just one example referring to this consensus, see D. W. Dunlap, "21st-century plans, but along 18th-century paths," *New York Times,* 17 July 2002. The plans for the rebuilding of the World Trade Center will create much welcome open space around the footprints of the old towers, but at this writing will not achieve mixed use to any significant degree.

17. Cluster zoning defined in Whyte 1964 and further in Whyte 1968.

18. Paul Stoller, consulting energy engineer, suggested this reference.

19. Living Machines Inc., www.livingmachines.com/htm/machine.htm.

20. The old World Trade Center had just about this same usable floor depth, except that in that case (as in most conventional buildings), this depth was from elevator core to windows.

References

Aristotle. *The politics.* 1253a1. London: Penguin Classics.

Augustine. 1958. *The city of God.* New York: Doubleday, Image Books (reprint).

Beatley, T. 2000. *Green urbanism: Learning from European cities.* Washington, DC: Island Press.

Brower, M., and W. Leon. 1999. *The consumer's guide to effective environmental choices: Practical advice from the Union of Concerned Scientists.* New York: Three Rivers Press.

Calthorpe, P. 1993. *The next American metropolis; ecology, community, and the American dream.* New York: Princeton Architectural Press.

Cathcart, C. M. 1998. SIPs not Studs. *Architecture* 6:148–52.

Davis, K. 1966. The urbanization of the human population. In *The city reader,* ed. R. LeGates and F. Stout, 1–14. New York: Routledge.

Duany, A., E. Plater-Zyberk, and J. Speck. 2000. *Suburban nation: The rise of sprawl and the decline of the American dream.* New York: North Point Press.

Gissen, D., ed. 2000. *Big and green: Toward sustainable architecture in the twenty-first century.* New York: Princeton Architectural Press / National Building Museum.

Ingersoll, R. 1996. Second nature: On the social bond of ecology and architecture. In *Reconstructing architecture: Critical discourses and social practices,* ed. T. A. Dutton and L. H. Mann, 119–57. Minneapolis: University of Minnesota Press.

Jacobs, J. 1961. *The death and life of great American cities.* New York: Vintage Books. Reissue ed., January 1993.

Kitto, H. D. F. 1951. *The Greeks.* Chicago: Penguin.

Koolhaas, R. 1979. *Delirious New York: A retroactive manifesto for Manhattan.* New York: Monacelli Press. Reprint ed., December 1997.

Le Corbusier.1925. The city of tomorrow and its planning. Originally published as *Urbanisme.* New York: Dover. Reissue ed., June 1987.

Light, A. 2001. The urban blind spot in environmental ethics. *Environmental Politics* 10(1): 7–35.

Lippe, P. 1998. *Earth Day New York, Lessons Learned: Four Times Square, An Environmental Information and Resource Guide for the Commercial Real Estate Industry.* New York: Earth Day New York.

Neuman, P. 1997. Greening the city: the ecological and human dimensions of the city can be part town planning. In *Eco-city dimensions: Healthy communities, healthy planet,* ed. M. Roseman, 14–24. Gabriola Island, British Columbia: New Society Publishers.

Newman, P., and J. R. Kenworthy. 1989. *Cities and automobile dependency: An international sourcebook.* Abingdon, UK: Gower Publishing.

Owen, D. 2004. Green Manhattan. *New Yorker* (18 October): 82–85; 120–23.

Palmer, C. 2003. An overview of environmental ethics. In *Environmental ethics,* ed. A. Light and H. Rolston III, xx. Malden, MA: Blackwell.

Plunz, R. 1990. *A history of housing in New York City* (Columbia History of Urban Life). New York: Columbia University Press.

Ponting, C. 1991. *Green history of the world.* New York: Penguin Books.

Rees, W. E., and M. Wackemagel. 1991. *Our ecological footprint.* New York: New Society Press.

Sullivan, L. 1896. The tall office building artistically considered. *Lippincott's.*

Watson, D. 2001. Environment and architecture. In *The discipline of architecture,* ed. A. Piotrowski and J. W. Robinson, 158–72. Minneapolis: University of Minnesota Press.

WCED [World Commission on Environment and Development]. 1987. *Our common future.* New York: United Nations.

Whyte, W. H. 1964. *Cluster development.* New York: American Conservation Association.

———. 1968. *The last landscape.* New York: Doubleday. Republished with a foreword by Tony Hiss, Philadelphia: University of Pennsylvania Press, 2002.

———. 1980. *The social life of small urban spaces.* Washington, DC: The Conservation Foundation.

Winner, D. 2002. *Brilliant orange, the neurotic genius of Dutch soccer.* Woodstock, NY: Overlook Press.

Yaro, R. D., and T. Hiss. 1996. *A region at risk: The third regional plan for the New York–New Jersey–Connecticut metropolitan area.* Washington, DC: Island Press.

Green Urbanism in European Cities

Timothy Beatley

In few other parts of the world is there as much interest in urban sustainability as in Europe, especially northern and northwestern Europe. Many European cities are pushing the envelope of urban sustainability, undertaking a variety of impressive actions, projects, and innovative policies to reduce their ecological footprints as well as to enhance long-term livability. For several years, I have been researching innovative sustainability practices in European cities, with many of the exemplary cases described in the book *Green Urbanism: Learning from European Cities* (Beatley 2000). What follows is a summary of some of the key themes from this research and most promising ideas and strategies found in these more than thirty cities, scattered across eleven countries.

Sustainability is an increasingly common goal at the local or municipal level in Europe and especially in the cities selected for study. The concept of sustainable cities or ecological cities resonates well at this local level and has important political meaning and significance in these cities and on the European urban scene in general. One measure is the success of the Sustainable Cities and Towns campaign, a European Union–funded informal network of communities pursuing sustainability begun in 1994. Participation has been great, with more than two thousand local and regional authorities having signed a sustainability charter (the so-called Aalborg Charter, after the Danish city that hosted the first campaign conference). Among the activities of this organization are the publication of a newsletter, networking between cities, and convening conferences and workshops. The organization has also created the European Sustainable City award, and it is clear that these awards have been coveted and highly valued by politicians and city officials. These European cities demonstrate serious commitment to environmental values and have much to teach about how to put them into practice.

European cities have also gone through or are going through extensive local Agenda 21 activities, typically resulting in the preparation of a local sustainability action plan and a host of tangible actions for making these communities more sustainable. These actions range from composing and recycling initiatives, to urban ecosystem restoration, to establishment of neighborhood sustainability centers. European city participation has been relatively high, with nearly 100 percent of municipalities participating in countries such as Sweden, for example. (For a review of Local Agenda 21 experience in Europe, see Lafferty 2001.)

Green Urbanism: Compact and Ecological Urban Form

Although European cities have become more decentralized, they are typically still more compact and dense than U.S. cities. This tighter urban form helps make local sustainability initiatives more feasible in terms of, for example, public transit, walkability, and energy efficiency. There are many factors that explain this urban form, including an historic pattern of compact villages and cities, a limited land base, and different cultural attitudes about land. Nevertheless, in the cities studied there are conscious policies aimed at strengthening a tight urban core. Indeed, the major new growth areas in almost every city studied are situated within or adjacent to existing developed areas and are designed at relatively high densities. Moreover, these new growth areas are incorporating a wide range of ecological design concepts, from solar energy to natural drainage to community gardens, and effectively demonstrate that *ecological* and *urban* can go together. Good examples of this *compact green* development can be seen in the new growth areas planned in Utrecht (Leidsche Rijn), Freiburg (Rieselfeld), Kronsberg (Hannover), Amsterdam (e.g., IJburg), Copenhagen (Ørestad), Helsinki (Viikki), and Stockholm (Hammerby Sjöstad). (See Beatley 2000 for further discussion of each area.)

Leidsche Rijn, a new growth district in Utrecht, incorporates a mixed-use design and a balance of jobs and housing (thirty thousand dwelling units and thirty thousand new jobs) as well as a number of ecological features. Much of the area will be heated through district heating supplied from the waste energy of a nearby power plant, a double-water system that will provide both potable and recycled water for nonpotable uses and stormwater management based on a system of natural swales (what the Dutch call *wadies*). Higher-density uses will be clustered around several new train stations, and bicycle-only bridges will provide fast, direct connections to the city center. Homes and buildings will meet a low energy standard and must use certified sustainably harvested wood. At Kronsberg, a host of green urban elements are integrated into this new ecological district, including three wind turbines, solar panels, district heating, onsite stormwater collection, green rooftops, and green courtyards and community gardens, all within a car-limited, pedestrian-friendly environment served by a new high-frequency tram line.

The new redevelopment of the Western Harbor (Västra Hamnen) in Malmö is another model example. Here, a former industrial area is being converted to a new living district, with sustainability as the key organizing principle. One of the main goals is to provide for 100 percent of the energy needs of the district from locally-generated renewable energy. Through the installation of a 2 megawatt wind turbine, and photovoltaics and solar hot water heating panels on building rooftops, this goal has already been achieved (see European Academy for the Urban Environment 2001). Other important ecological elements include a circular waste treatment system in which biogas is extracted from organic waste and returned to

University Physicians
Dept. 99
UConn Health Center

DATE _4-1-08_

RECEIVED
FROM _Jack Hale_

FOR _____

_T0021241_

DATE OF SERVICE	☒ CASH	PHYSICIAN NAME
4-1-08	☐ CHECK	Taylor

C V Copay

_57/14414_

DOLLARS		RECEIVED BY:
20	00	VJM
	CENTS	

REMARKS _____

_____

787220

the district through the natural gas grid, on-site collection of rainwater (and creative urban design that marvelously integrates water into the district and makes it visible throughout), and extensive natural habitat creation (figure 1). These new developments show convincingly that *green* and *urban* go together, indeed are complementary and mutually reinforcing, creating compelling and highly livable communities that exert an impressively small demand on the earth's resources.

These cities also provide examples of redevelopment and adaptive reuse of older, deteriorated areas within or near the center city. In Amsterdam's eastern docklands, eight thousand new homes have been constructed on recycled land. In one part of this project, *Java-eiland*, design diversity has been encouraged through the use of multiple architects. The overall plan for this island district successfully balances connection to the past (a series of canals and building scale reminiscent of historic Amsterdam) with unique modern design (each of the pedestrian bridges crossing the canals offers a distinctive look). *Java-eiland* demonstrates that city building can occur in ways that create interesting and organically evolved places and that also acknowledge and respect history and context, and overcome monotony.

One of the boldest ecological restoration and land recycling initiatives has taken place in the industrial Ruhr Valley of northwestern Germany, consisting of former coal mines and steel mills. Here a regional regeneration strategy has been implemented, including seventeen municipalities and an urban agglomeration of two million people. The bold effort involved formation of IBA-Emscher Park, an international exhibition, comprising some 120 different reuse projects over an eight-hundred-square-kilometer area. The projects range from the conversion of a large gasometer to exhibition space, to transforming slag heaps into parks and public art. In the process, these bold initiatives have fundamentally reshaped the local perception of this formerly bleak, industrial landscape. One spectacular example is the Duisberg-Nord Landscape Park, where a former steel mill has been miraculously transformed into a unique city park (figure 2). Formal gardens have been carved out of coal and coke storage areas, foundation walls are turned into climbing and repelling areas, and the blast furnace a kind of industrial Eiffel Tower. Here visitors "cannot help but be awed by the skill and strength demanded of the men who once produced iron and steel here" (LaBelle 2001, 225).

It is an odd landscape of "industrial monuments" and landscape art, the latter converting negative remnants of the industrial landscape into a most interesting and positive aesthetic. As Judith LaBelle notes, the art was important for signaling a new direction:

> The art has helped to signal the forward-looking nature of the initiative and to provide a system of new landmarks through the landscape. Several large sculptures have been installed atop slag heaps, including the towering Tetrahedron at Bottrop. Lighted at night, they provide new reference points in the night landscape. Smaller, more intimate sculptures have been created in areas newly used for parks and recreation. They serve to draw

Figure 1 Vertical solar hot water panels in the sustainable planned district Vastra Hamnen, in Malmö, Sweden. (Photo by Tim Beatley.)

Figure 2 The Landscape Park in Duidberg-Nord in Germany's Ruhr Valley includes creative reuse of a former steel mill. (Photo by Tim Beatley.)

this visitor into a landscape that has hitherto been off-limits and foreign. Some are composed of industrial artifacts found on the site, providing a more intimate connection with the site's history. (LaBelle 2001, 226)

These European cities have also undertaken numerous efforts to enhance the quality and attractiveness of their city centers. In the cities studied, the center has remained a mixed-use zone, with a significant residential population. *Groningen,* for instance, has created new pedestrian-only shopping areas (a system of two linked circles of pedestrian zones) and has installed yellow brick surfaces and new street furniture. Committed to a policy of compact urban form, Groningen has made a strong effort to keep all major new public buildings and public attractions close to the center. A new modern art museum in that city has been sited and designed to provide an important pedestrian link between the city's main train station and the town center.

Freiburg has done much to improve the attractiveness of its center: gradually making much of the core more available to pedestrians, maintaining housing and people living in the center (e.g., forbidding the conversion of existing downtown housing to commercial and other uses), and strengthening the visual landmarks and aesthetic qualities of the old city center. Especially unique is the city's network of small water channels in the city's streets (so-called *bächle*), which add a special flavor and enjoyable quality to this place. Developers of new projects in the city are now asked to build onto and expand this unique system, furthering strengthening these unique and special qualities.

Many reasons help explain why these European cities are able to achieve a more compact urban form. In countries such as the Netherlands, there are clearly stronger public planning systems in place, with a considerably greater role for provincial and national governments (e.g. see Van den Brink and Van der Valk, 2002). A generally greater public-sector role in shaping development and growth, restrictions on private land use, and economic incentives that encourage cooperation and more sustainable outcomes (e.g., much higher gasoline and energy prices, carbon taxes) is also a significant factor. A different attitude about land—one that views it as a precious and limited resource—and a cultural affinity for urban settlements and living are, to be sure, also important factors.

Sustainable Mobility

Achieving a more sustainable mix of mobility options is a major challenge, and in almost all the cities studied in *Green Urbanism,* a very high level of priority is given to building and maintaining a fast, comfortable, and reliable system of public transport. Zürich, for instance, gives priority to its transit on streets with dedicated lanes for trams and buses and numerous improvements to reduce the interference of autos with transit. A single ticket is good for all modes of transit—including

Figure 3   A pedestrian, car-free center in Freiburg, Germany. (Photo by Tim Beatley.)

buses, trams, and a new underground regional metro system—in the city. The frequency of service is high and the spatial coverage extensive. Cities like Freiburg and Copenhagen have made similar strides (figure 3).

Commitment to excellent public transit services is a hallmark of many European cities. Cities like Zürich, where public transit has been given official priority over cars, seek to make transit faster, easier, and more pleasant to ride and to coordinate service extensions with new housing. Integrated transport systems, where movement from one mode to another is made easy and where riders have real-time information on when the next train, tram, or bus will arrive, are consistent qualities. (See Newman and Kenworthy 1999 for a good discussion of European public transit systems.)

In most of the cities studied, regional and national trains systems are fully integrated with local routes. It is easy to shift from one mode to another. And, with the continuing commitment to the development of a European high-speed rail network, modal integration is becoming even greater.

Furthermore, transit investments complement, and are coordinated with, important land use decisions. The new development *Rieselfeld* in Freiburg, for instance, has a new tram line even before the project has been fully built. In Amsterdam's new neighborhood of *Nieuw Sloten,* tram service began when the first homes were built. In the new ecological housing district *Kronsberg*, in Hannover, three new tram stops ensure that no resident is farther than six hundred meters away from a station. There is a recognition in these cities of the importance of providing new residents with options and establishing mobility patterns early.

Car sharing has become a viable and increasingly popular option in Europe cities. Here, by joining a car-sharing company or organization residents have access to neighborhood-based cars on an hourly or per kilometer cost. There are now more than one hundred thousand members served by car-sharing companies or organizations in more than five hundred European cities. Some of the newest car-sharing companies, such as *GreenWheels* in the Netherlands, are also pursuing some creative strategies for enticing new customers. GreenWheels has been developing strategic alliances, for example with the national train company, to provide packages of benefits at reduced prices. A key issue for the success of car-sharing is the availability of convenient parking spaces, and a number of cities, including Amsterdam and Utrecht, set aside spaces for this purpose. In cities such as Hannover, Germany, the car-sharing organization there (a nonprofit called Okostadt) has strategically placed cars at the stations of the Stadtbahn, or city tram, further enhancing their accessibility.

Taming the Automobile

Many of these cities are on the vanguard of new mobility ideas and concepts and are working hard to incorporate them into new development projects. Amsterdam, for example, has taken an important strategy in developing *IJburg*, its newest growth area. It is working to develop a comprehensive mobility package that all new residents will be offered and that include, among other things, a free transit pass (for certain specified period) and discounted membership in local car-sharing companies. Minimizing from the beginning the reliance on automobiles and giving residents more mobility options are the goals. Eventually this new area will be served both by an extension of the city's underground metro and fast tram (Beatley 2000).

An increasing number of car-free housing estates are being developed to further reduce auto dependence. The *GWL-Terrein* project, built on Amsterdam's old waterworks site, incorporates only limited peripheral parking. Mobility is assisted by good tram service and, when a car is needed, an on-site car-sharing company. The interior of the project incorporates extensive gardens (with 120 community gardens available to residents) and a pedestrian-friendly environment, with key-lock access for fire and emergency vehicles.

Freiburg's *Vauban*, another car-free district, charges residents approximately $18,000 for the cost of a space in the local parking garage (about one-tenth the cost of the housing units), a strong disincentive to car ownership. Projects like *Vauban* challenge new residents to think and act more sustainably (figure 4).

Figure 4 In Freiburg, Germany, new housing districts such as Vauban, pictured here, are designed to discourage car ownership and use. (Photo by Tim Beatley.)

Many of the cities studied have made tremendous efforts to expand bicycles facilities and promote bicycle use. Berlin has eight hundred kilometers of bike lanes, and Vienna has more than doubled its bicycle network since the late 1980s. Many actions have been taken by these cities to promote bicycle use, such as separated bike lanes with separate signaling, priority at intersections, signage, and provision of extensive bicycle parking facilities, including minimum bicycle storage and parking standards for new development.

Some cities actively promote "public bikes." The most impressive is Copenhagen's *City Bikes program,* which makes available some two thousand public bicycles throughout the center of the city. The bikes are brightly painted (companies sponsor and purchase the bikes in exchange for the chance to advertise on their wheels and frames) and can be used by simply inserting a coin as a deposit. The bikes are geared in such a way that the pedaling is difficult enough to discourage their theft. The program has been a success, with the number of bikes increasing. More recent have been efforts at developing higher-tech systems of "smart bikes." For instance, Deutsche Bahn, the German train company, has been experimenting with a system of bikes available at major train stations, such as Frankfurt. The bikes can be reserved by phone or on line and can be accessed through electronic locking pads installed on the bikes. These bikes are easily dropped off at one of a number of points around the city, and a rider's credit card is charged for the actual time the bike is used (figure 5).

Greening the Urban Environment

Ensuring that compact cities are also green cities is a major challenge, and there are a number of impressive greening initiatives among the study cities. First, in many of these cities there is an extensive greenbelt and regional open space structure, with a considerable amounts forest and natural land owned by cities such as Vienna, Berlin, and Graz. Cities such as Helsinki and Copenhagen are spatially structured so that large wedges of green nearly penetrate the center for these cities. Helsinki's large *Keskuspuisto* central park extends in an almost unbroken wedge from the center to an area of old growth forest to the north of city, one thousand hectares large and eleven kilometers long.

Hannover boasts an extensive system of protected greenspaces, including the *Eilenriede*, a 650 hectare dense forest located in the center of the city. Hannover has also recently completed an eight-kilometer-long green ring *(der Grune Ring)* that circles the city, providing a continuous hiking and biking route and exposing residents to a variety of landscape types.

Ecological networks are being developed within and between urban centers. They are perhaps most evident in Dutch cities, where extensive attention to ecological networks has occurred at the national and provincial levels. Under the national government's Nature Policy Plan, a national ecological network has been established consisting of core areas, nature development areas, and corridors, which must be more specifically elaborated and delineated at the provincial level. In turn, cities are attempting to tie into this network and build upon it.

Greening initiatives may be mandated or subsidized by public authorities. German, Austrian, and Dutch cities are especially proactive concerning ecological or green roofs. Linz, Austria, for instance, has one of the most extensive green roof programs in Europe. Under this program, the city frequently requires building plans to compensate for the loss of greenspace taken by a building, resulting in the creation of many green roofs. The city also will pay up to 35 percent of the costs and provide technical assistance to facilitate green roofs. Some three hundred green roofs are now scattered around the city atop many kinds of buildings, including a hospital, a kindergarten, a hotel, and even a gas station. Green roofs have been shown to provide a number of important environmental benefits and to accommodate a surprising amount of biological diversity (figure 6). Many other innovative urban greening strategies can be found in these cities, from green streets to green bridges to urban stream daylighting (see Beatley 2004).

Figure 5 The Deutsche Bahn, Germany's national train system, has been experimenting with a unique system of "smart bikes" at train stations equipped with electronic locking pads that can be reserved in advance. (Photo by Tim Beatley.)

Figure 6   Green or ecological rooftops are standard design features in European cities, such as this example in central Amsterdam. (Photo by Tim Beatley.)

Renewable Energy and Closed-Loop Cities

A number of the cities seek to promote a more closed-loop or natural urban metabolism in which wastes become inputs or food for other urban processes. Stockholm has administratively reorganized its departments of waste, water, and energy into a combined ecocycles division. A number of actions have already been taken, including the harvesting of bio-gas from sewage sludge and its use as a fuel for the city's combined heat and power plants. A number of Swedish cities also are using bio-gas from household waste as a fuel for buses and other public vehicles (Swedish Ministry of the Environment, undated; for a review of environmental vehicle programs in European cities, see European Commission 2001). Experience to date suggests that in addition to recycling waste there has been a dramatic reduction in conventional air pollutants as well as in carbon dioxide emissions in these cities. Another powerful example of the closed-loop concept can be seen in Rotterdam's Roca3 power plant, which supplies district heating and carbon dioxide to 120 greenhouses in the area (figure 7). A waste product becomes a useful input and in this case prevents some 130,000 metric tonnes of carbon emissions annually.

Important strategies in northern European cities are combined heat and power generation and district heating. More than 91 percent of Helsinki's buildings, for instance, are connected to district heating systems, resulting in a substantial increase in fuel efficiency and reductions in pollution emissions. In *Kronsberg,* in Hannover, heat is provided by two combined heat and power plants, one of which, serving about six hundred housing units and a school, is actually located in the basement of a building of flats (see Landeshaupt Stadt Hannover 2000; Beatley 2004).

Heidelberg and Freiburg, among other cities, have set ambitious maximum energy consumption standards for new construction projects. Heidelberg has recently sponsored a low-energy social housing project to demonstrate the feasibility of very low energy designs (specifically, a standard of 47 kilowatt-hours per square meter per year). The Dutch have been promoting the concept of "energy-balanced housing"—homes that produce at least as much energy as it uses over the course of a year—and have built them in new development areas such as Nieuwland in Amersfoort (see figure 8). In Leeuwarden in the Friesland region of the Netherlands, Europe's first energy-balanced street has been completed. More recently,

Figure 7 The Roca-III power plant in Rotterdam supplies heating and carbon dioxide to some 120 greenhouses in its vicinity. (Photo by Tim Beatley.)

Stad von de Zon (City of the Sun), a new community north of Amsterdam, has been designed to be both energy balanced and carbon neutral.

Solar energy and other renewable energy sources are of great importance at the city or municipal level, and there is increasingly the view that cities must lead the way in charting a new path beyond and away from fossil fuels. Cities like Freiburg and Berlin have been competing to be know around Europe and the world as "solar cities," with each providing significant subsidies for solar installations. In the Netherlands, major new development areas have incorporated both passive and active solar energy. In Nieuwland, nine hundred homes have rooftop photovoltaics, eleven hundred homes have thermal solar units, and a number of major public buildings (including several schools, a major sports hall, and a child care facility) produce power from solar energy.

Green Cities, Green Governance

Many of the cities studied seek to set their own environmental houses before asking their citizens to act more sustainably. Some are looking at how their own operations and management can be more environmentally responsible. Albertslund, Denmark, for example, has developed an innovative system of "green accounts" used to track and evaluate key environmental trends at city and district levels. Den Haag has calculated the average ecological footprint of its residents and begun using it as a policy guidepost. Several German cities are using ecological budgets alongside their conventional fiscal budgets, and Rome has been developing a similar system of environmental accounts. The Swedish city of Sundesvall has since 1991 published an annual "environmental balance sheet" (or Mijöbokslut), which takes stock of current environmental conditions in the city and actions taking over the course of the year (UBC 2002).

Municipal governments have taken a variety of measures to reduce the environmental impacts of their actions. A number of communities have adopted environmental purchasing and procurement policies. Alberstlund, Denmark, mandates that only organic food can be served in schools and child care facilities, and the city restricts the use of pesticides in public parks and grounds. Several cities promote the use of environmental vehicles, and some, like Saarbrucken, Germany, have made great progress in reducing energy, waste, and resource consumption in public buildings.

Some communities have engaged in extensive community involvement and outreach on sustainability. Leicester, England, for instance, has developed alliances with the local media and has sponsored a series of educational campaigns on particular community issues. As a further example, it also runs a demonstration ecological home and garden. Cities like Albertslund have opened neighborhood environmental centers as an effective way to engage and educate the community.

Some Lessons and Observations

European cities thus offer inspiration and lessons for cities elsewhere, including the United States. A few observations and lessons follow.

Government as catalyst and leader. European cities display a strong role for municipal governments in shaping sustainable futures. They tend to assume activist and catalytic roles in diverse ways. For instance, they exert considerable control over the use of land and the type, quality, and nature of private development. They typically acquire or already own the land for large new housing areas, prepare detailed plans, install the infrastructure, and establish very specific contractual requirements for builders and developers to follow.

They often use the city's purchasing power to support sustainable technologies, to educate consumers, and to help local businesses become more sustainable. Sustainable technologies are commonly subsidized and underwritten by cities as, for instance, grants for the installation of green rooftops. Thus, these cities actively support and promote a vision of a more ecological or humane urban community.

Pushing the ecological design envelope. Many European cities are promoting green technology and new ecological living ideas on an unprecedented scale. New urban districts like Leidsche Rijn and IJburg in the Netherlands are applying green urban ideas to thousands of new homes. New solar projects like the Stad van de Zon, in Heerhugowaard, are aspiring to be carbon neutral, and projects like the Western Harbor in Malmö are already achieving the goal of 100 percent locally produced, renewable energy. They are bold goals and visionary plans, indeed, for how to craft humane, sustainable places for our future.

Comprehensive green strategies. European cities treat sustainability comprehensively. Cities like Freiburg are simultaneously implementing programs to promote solar energy, walking, bicycling and transit use, car-free living, and ecological landscape management. Such green initiatives tend reinforce each other. Strengthening public transit and pedestrian and bicycle use undergirds car-free housing development. Green building and ecological regeneration may help stabilize neighborhoods and reduce turnover in social housing. Thus, one green urban policy can strengthen and complement other social objectives. Moreover, every major building project in these cities is viewed as a chance to promote experimentation, to set and reach new ecological goals, and to demonstrate the integration and application of new ecological ideas and technologies.

New ways of seeing cities. Cities and city life are viewed in new ways in Europe. Rooftops of sports halls and schools in Nieuwland, in Amersfoort, are viewed as opportunities to generate power as well as opportunities to educate children and the community about energy issues. Cities are viewed not simply as points of consumption but as places where renewable energy can be produced (and consumed) and integrated into the built fabric. Many of these cities are redefining themselves in terms of a circular metabolism: the Swedish refer to this as ecocycle balancing (Girardet 1999; Rogers 2000).

The perspective of cities as places of nature and as organic natural systems is also taking hold. Cities need not be opposed to or in conflict with nature. Rather, they can and should be seen as inherently embedded in a natural system and condition. Nature in cities is being enhanced by such features as ecological rooftops, green streets, and stream daylighting.

Changing economic incentives. On many levels and in many ways, Europeans recognize the importance of leveling the economic playing field to support green urban ideas and technologies. Such leveling takes many forms. Many cities charge homeowners for the extent of impervious surfaces while reducing stormwater fees for homes with green rooftops and permeable driveways, for instance. Subsidizing and investing heavily in nonautomobile infrastructure as well as charging (closer to) the full cost of auto ownership and use (e.g., the experience of the Vauban car-free housing estate in Freiburg) are also examples of this philosophy. Subsidies for green projects and practices often work in tandem with stronger regulations in these European countries. Barcelona's municipal solar ordinance, for instance, mandates that solar panels provide at least 60 percent of the hot water needs of new and renovated buildings, but at the same time the city provides significant subsidies to encourage installation of solar panels.

The importance of networks. Cities are not only operating in different ways within their borders; they are also operating in creative ways among and between themselves. Especially impressive is the extensive use of networks and association of cities. The Sustainable Cities and Towns campaign is an excellent example. On a smaller scale, the Union of Baltic cities (UBC) provides a similar technical and peer support function. Today, more than one hundred Baltic cities participate in the UBC. Through the UBC, meetings, workshops, and seminars are convened, and municipalities share information and insights and provide mutual support. An initiative called the Best City Practices Project, as one example, has paired Baltic cities together in an exchange of knowledge and experience on sustainable development issues (see UBC 2002). A Best Environmental Practice in Baltic Cities Award is also given each year to support good ideas and practice.

Many European cities are facing serious problems and trends working against sustainability: a dramatic rise in auto ownership and use and a continuing pattern of deconcentration of people and commerce. European cities also exert a tremendous ecological footprint on the world. Yet these most exemplary cities provide both tangible examples of sustainable practice and inspiration that progress can be made in the face of these difficult pressures.

Moreover, these examples demonstrate the critical role that cities can and must play in addressing the serious global environment problems, including overreliance on fossil fuels and global climate change. Innovations in the urban environment offer tremendous potential for dramatically reducing our ecological impacts, while at the same time enhancing our quality of life (e.g., expanding personal mobility options with bicycles and transit). The experiences demonstrate clearly that it is possible to apply virtually every green or ecological strategy or technique—from solar and wind energy to gray-water recycling—in very urban, very compact settings. *Green urbanism is not an oxymoron.* Moreover, the lesson of these European cities is that municipal governments can do much to help bring these ideas about, from making parking spaces available for car-sharing companies to providing density bonuses for green rooftops to producing or purchasing green power.

It is important to recognize that the lessons are not just in one direction. Increasingly, European cities recognize that there are aspects of U.S. planning and policy that are helpful and can provide useful lessons for them as well. Ari Van den Brink and Arnold Van der Valk (2002), for instance, argue that European planning systems have historically tended to be more technocratic (more top down, with greater power given to plans and planners), whereas the U.S. system is more "sociocratic" (bottom up, participatory, deferential to the wishes of individuals). European planners recognize the need to be more participatory. Techniques such as community visioning tools and citizen design charrettes represent ideas that Europeans increasingly find useful and interesting. There is much, then, to share in both directions.

References

Beatley, T. 2000. *Green urbanism: Learning from European cities.* Washington, DC: Island Press.

———. 2004. *Native to nowhere: Sustaining home and community in a global age.* Washington, DC: Island Press.

European Academy of the Urban Environment. 2001. Mälmo: Bo02, city of tomorrow: European building exhibition and sustainable development. EAUE good practices database. Available online at www.eaue.de.

European Commission. 2001. *Cleaner vehicles in cities: Guidelines for local governments.* Utopia Project, February. Available online at http://utopia.jrc/d19.

Girardet, H. 1999. *Creating sustainable cities.* Devon, UK: Green Books.

LaBelle, J. M. 2001. Emscher Park, Germany—expanding the definition of a park. In *Crossing boundaries in park management: Proceedings of the Eleventh Conference on Research and Resource Management in Parks and on Public Lands,* ed. D. Harmon. Hancock, MI: The George Wright Society.

Lafferty, W. M., 2001. *Sustainable European communities.* London: Earthscan Books.

Landeshaupt Stadt Hannover. 2002. *Modell Kronsberg: Sustainable building for the future.* Hannover, September.

Newman, P., and J. Kenworthy. 1999. *Sustainable cities.* Washington, DC: Island Press.

Rogers, R. 2000. *Cities for a small planet.* Boulder, CO: Westview Press.

Swedish Ministry of the Environment. Undated. Local climate protection measures in Sweden—best practices within the local investment programmes. Stockholm: Ministry of the Environment.

UBC [Union of Baltic Cities]. 2002. *Best city practices report.* Turku, Finland: UBC.

van den Brink, A., and A. van den Volk. 2002. Digging the Dutch: Planning principles and practice in the delta metropolis. Paper presented to the Fourth International Workshop on Sustainable Land Use, Bellingham, WA, June.

Epilogue

Pathways to More Humane Urban Places

Rutherford H. Platt

William H. Whyte's 1957 essay "Urban Sprawl" was indeed prescient: despite the open space movement of the 1960s (which he helped to nurture) and its outgrowths—growth management, smart growth, and New Urbanism—metropolitan expansion has continued relentlessly. In 1961, geographer Jean Gottmann defined "Megalopolis" as a region of more or less continuous urbanization extending along the northeastern seaboard from just north of Boston to the Virginia suburbs of Washington, D.C. Since then, Megalopolis has sprawled north, west, and south beyond its 1960s geographic size.

Megalopolis today would include southeastern New Hampshire and the southern Maine coast, Massachusetts west to the Berkshires, the Hudson River Valley north to Lake George, much of New Jersey and eastern Pennsylvania, most of Maryland, portions of West Virginia, and the I-95 corridor south at least to Richmond, Virginia, a vast megaregion covering parts of thirteen states and containing nearly fifty million people. Just southwest of that, a new complex following I-85 and I-40 connects the North Carolina metro areas of Charlotte, Greensboro, and Raleigh–Durham–Research Triangle. Greater Atlanta now reaches more than 110 miles north to south, compared with 65 miles in 1990 (Bullard, Johnson, and Torres 2000, 9). Both coasts of Florida are solidly lined with metropolitan areas. Greater Chicago extends well into northwestern Indiana and southeastern Wisconsin. The Colorado "Front Range urban corridor" reaches from Pueblo northward to Fort Collins and Greeley, encompassing metropolitan Denver, Colorado Springs, and Boulder. Greater Los Angeles is spilling eastward across the "inland empire" of Riverside and San Bernadino counties into the Mojave Desert. Irrigated farms of California's Central Valley are disappearing under pavement, and the fringes of Portland and Seattle are flirting with each other along the foothills of the Cascades. (These trends have been recently analyzed by Robert E. Lang and Dawn Dhavale (2005a, 2005b).

As discussed in the introduction of this book, metropolitan areas (central cities plus their suburbs) nearly doubled in population, from 118 million in 1960 to 226 million in 2000: *80 percent* of Americans now live and work in metropolitan areas, which themselves have about doubled in total geographic area since the 1960s. Perversely, the fastest population growth has occurred where nature is least welcoming: the desert settings of metropolitan Las Vegas (1990–2000 population increase of 83.3 percent), Phoenix (+45.3 percent), Tucson (+26.5 percent), and Riverside–San Bernardino (+25.7 percent).

Furthermore, the rate of land consumption has far outpaced the rate of population growth for most metropolitan areas: metro Los Angeles expanded by 300 percent in urbanized land area between 1970 and 1990, while its population grew by only 45 percent; the Seattle region grew by 38 percent in population, but 87 percent in urbanized area (see table 2 in the introduction). Even metropolitan Pittsburgh expanded 43 percent in urbanized land area between 1982 and 1997 despite a regional *population decline* of 8 percent during the same period (Sustainable Pittsburgh 2003, 3). Metropolitan Portland, Oregon, however, experienced only a 4 percent increase in developed area despite a 31 percent increase in population between 1990 and 2000, owing largely to its urban growth boundary, which limits sprawl onto surrounding farmland and forestland (Michael Houck, personal communication, May 19, 2005).

Sprawl has been further exacerbated by the trend toward ever-larger single-family homes and lots on the suburban fringe. Average floor area per capita in new single-family homes has tripled since the 1950s, and average lot sizes have grown correspondingly (Brewster 1997, 7).

As urban sprawl has enveloped ever more of the nation's population and accessible land area, perception of its harmful impacts—on society, the economy, and the environment—has broadened as well. The early critiques by William H. Whyte, Charles Little, Ann Louise Strong, and others in the 1960s focused primarily on issues of *aesthetics* and *efficient land use* related to urban encroachment on productive farmland and the loss of access to "countryside." To those still valid concerns have been added a variety of further concerns including *air and water pollution, waste of energy and time, traffic congestion and highway accidents, lack of affordable housing, "brownfields," water scarcity, increased flooding, and loss of biodiversity* (Gillham 2002, 75–77).

Beyond such direct consequences are secondary sets of implications, such as (1) the *fiscal burdens* of providing infrastructure and public services to fringe development (Diamond and Noonan 1996, 34–40); (2) *emotional stress* on individuals and families due to separation of home, workplace, and other destinations; (3) loss of *sense of community* (Putnam 2000); and (4) *social and environmental justice issues,* such as unequal access to housing, jobs, schools, and health services and exposure to environmental hazards. Moreover, sprawl itself with all its social inequity is a product of deliberate public policies concerning taxation, transportation, and local zoning (Bullard, Johnson, and Torres 2000; Platt 2004a, 2004b).

Historically, it has been an American tradition to leave place-based problems behind and seek "greener pastures" through relocation: to the frontier, to the suburbs, to the Sunbelt, and to the coasts, mountains, and deserts. In the process, however, the metropolis has often been an unwelcome hitchhiker. Metropolitan conditions have spread to such traditional vacation and retirement meccas as Cape Cod, the Maryland Eastern Shore, the Outer Banks of North Carolina, the Sierra and Rocky Mountain foothills, and the golf course utopias of the Southwest. As the urban fringe recedes indefinitely in travel time and distance, once treasured destinations increasingly resemble what people are trying to escape: traffic congestion, billboards, shopping malls, and general roadside schlock—"The Exploding Metropolis" writ large. Meanwhile, the less fortunate, the unemployed, the infirm, and the elderly are sentenced to live and die in the metropolitan environment, come what may. As Lewis Mumford wryly observed, "The ultimate effect of the suburban escape in our time is, ironically, a low-grade uniform environment from which escape is impossible" (Mumford 1961, 486).

This book and the conference from which it arose take a more upbeat look at the evolving form and substance of twenty-first-century metropolitan America. The term *humane metropolis* was chosen deliberately as a counterpoint to Whyte's "exploding metropolis" of the 1950s. The metropolis has indeed "exploded," most of us live in it, and so what are we going to do to make it more habitable? The humane metropolis was defined in the introduction as an urban region that is *more green, safer and healthier, more people friendly, and more socially equitable.* In the spirit of Holly Whyte, Jane Jacobs, and other "people who like cities," including the editor and authors of this book, we have explored diverse pathways to more humane urban places.

The primary "pathways" to more humane metropolitan regions are reflected in major sections of this book: Part I, "The Man Who Loved Cities"; Part II, From City Parks to Urban Biosphere Reserves; Part III, Restoring Urban Nature: Projects and Process; Part IV, A More Humane Metropolis for Whom?; and Part V, Designing a More Humane Metropolis. Certain essays directly relate to Whyte's own interests, such as the design of city and regional open space systems (Harnik, Houck), public attachment to city parks (Ryan), the smile index (Wiley-Schwartz), and the use of zoning incentives to create public spaces (Kayden). Other chapters, however, discuss twenty-first-century dimensions of the humane metropolis that we assume Whyte would embrace today, including social and environmental equity (Blakely, Anthony, Parrilla, Popper and Popper), regreening of brownfields (De Sousa) and ecological rehabilitation of closed landfills (Clemants and Handel), green building design (Pelletier, Cathcart), urban watershed management (Sievert), and the idea of "ecological citizenship" (Light 2002).

Several premises underlie and connect the various topics discussed in this book:

1. Most Americans now live and work in metropolitan regions.
2. Contact with, and awareness of, "nature" is a fundamental human need.

3. Access to unspoiled "nature" beyond metro areas is increasingly limited by distance, cost, traffic congestion, and tourist/resort development.
4. "Urban ecology" is not an oxymoron; nature abounds in urban places, if you know where and how to find it.
5. Therefore, opportunities to experience nature *within* urban places must be protected and enhanced.
6. Furthermore, protecting and restoring "ecological services" is often preferable to using technological substitutes.
7. Environmental education for all ages is critical to build support for such programs and to nurture a sense of "ecological citizenship."

The last three of these premises are critical to adapting to the twenty-first-century metropolis (Platt 2004b). Even as urban design professionals continue to manipulate the physical form and appearance of the *built* environment, new approaches, including some described in this book, focus on the *unbuilt* elements of the urban environment. Such adjustments are concerned less with the way urban places "look" and more with the way they "work," ecologically and socially.

The recognition that cities and nature are symbiotic rather than oxymoronic was long retarded by the professional disdain of natural scientists for cities. For instance, an influential Conservation Foundation book of the mid-1960s, *Future Environments of North America* (Darling and Milton 1965), virtually ignored urban places even though they were the "future environments" of most North Americans. As recently as 1988, a prominent National Academy of Sciences book titled *Biodiversity* (Wilson 1988) devoted a mere seven of 520 pages to "urban biodiversity." The view of nature as "out there" beyond the urban fringe or in exotic and distant places accessible only to scientists and the affluent ecotourist has often been reinforced by well-meaning natural history museums, zoos, aquaria, and television nature documentaries.

The seed of a different perspective on cities and nature was planted by landscape architect Ian McHarg in his seminal 1968 book *Design with Nature*. McHarg urged urban designers to evaluate and incorporate natural factors such as topography, drainage, natural hazards, and microclimate into their plans, rather than overcoming such constraints through technology–, often at great cost and with uneven success. McHarg's advice was directed primarily to the planning of new and often upscale suburban development. The proposition, however, would be significantly expanded by Anne Whiston Spirn in her book *The Granite Garden:* "The city, suburbs, and the countryside must be viewed as a single, evolving system within nature, as must every individual park and building within that larger whole. . . . Nature in the city must be cultivated, like a garden, rather than ignored or subdued" (1985, 5). In 1987, *The Greening of the Cities* examined British experience with "cultivating nature in cities," proposing that ecology offers "a way out of manmade aesthetics and proprietorial landscapes" (Nicholson-Lord 1987, 115). In a more emotional voice, evolutionary biologist Lynn Margulis and her son, Dorion Sagan, put it this way: "The arrogant habitat-holocaust of today may cease; in its wake may evolve technologically nurtured habitats that re-bind, re-integrate, and re-merge us with nature" (Margulis and Sagan 1999, 1).

Various terms today encompass efforts to regreen cities: *green urbanism* (Beatley 2000), *green infrastructure, natural cities* (Lord, Strauss, and Toffler 2003), variations of *urban sustainability,* and my own preference, *ecological cities* (Platt, Rowntree, and Muick 1994). Whatever the term, such approaches are typically localized, practical, and diverse. According to planner Timothy Beatley, green urbanism in European cities includes such elements as green roofs, community gardens, car-free neighborhoods, pavement removal, passive solar heating, and cohousing. Many of these are beginning to appear in U.S. cities at various scales and encompassing a broad spectrum of goals and means, as depicted in figure 1.

Some strategies that have been identified by the Ecological Cities Project (www .ecologicalcities.org), based at the University of Massachusetts Amherst, include the following:

- Rehabilitation and adaptation of older parks and urban green spaces
- Protection and restoration of urban wetlands and other sensitive habitat
- Preservation of old-growth trees and forest tracts
- Development of greenways and rail trails
- Urban gardening and farm markets
- Green design of buildings, including green roofs and green schools
- Brownfield remediation and reuse
- Urban watershed management
- Riverine and coastal floodplain management
- Endangered species habitat conservation plans
- Urban environmental education sites and programs
- Environmental justice programs.

Such efforts are typically initiated by nongovernmental organizations (NGO's) such as museums and botanic gardens, schools and colleges, watershed alliances, and regional chapters of national organizations like The Nature Conservancy, Trust for Public Land, Sierra Club, and National Audubon Society. NGOs provide vision, persistence, and sometimes volunteers to work in the field. Public-sector agencies play supporting roles: funding, staff resources, technical know-how, and (when applicable) regulatory muscle. Funds also may be contributed by businesses, foundations, and individuals, especially for projects in localities of particular interest to the donor (as with the Heinz and Mellon foundations in the Pittsburgh area and the Rockefeller Brothers Fund in New York). Researchers in universities, public agencies, and NGOs help define the scientific and social goals and means.

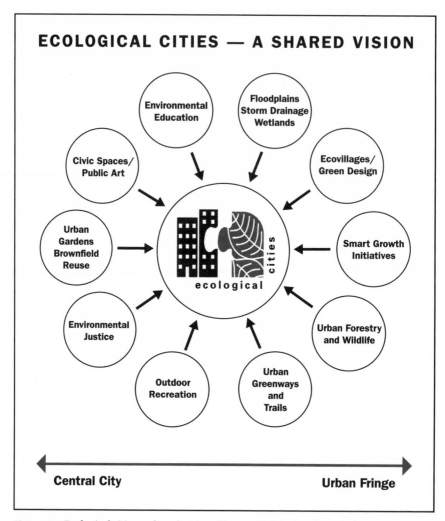

Figure 1   Ecological cities: a shared vision. (Source: University of Massachusetts Ecological Cities Project.)

Urban regreening efforts are often scattered, uneven, and underfunded, but like ecological organisms, they thrive on diversity: of goals, of means, of participants, of disciplines, and (one hopes) of viewpoints. Some are closely related to larger national movements such as social and environmental justice, affordable housing, physical fitness, public health, natural disaster mitigation, animal rights, and environmentalism. Some depend on spontaneous and often voluntary local leadership. They are pragmatic and creative in stitching together existing program resources, available funding, and donations of money, time, and office space. Most involve public-private partnerships, some of which are local alliances to save a particular site, to restore a stream, wetland, or watershed, or to pursue a particular mission such as environmental education or urban gardening. Others have evolved into influential regional networks such as Chicago Wilderness. Many also foster social interaction among diverse populations sharing a common resource like a watershed, thus promoting ecological citizenship. (See Light's essay in this volume.)

The half-century between the exploding metropolis and the humane metropolis thus spanned a period of vast change in the size, distribution, and habitability of urban places and regions. Although the negative implications of rampant urban growth have been widely deplored, efforts to curb the outward expansion of metropolitan areas have been largely futile. In the decades ahead, the emphasis must shift from limiting "urban sprawl" to making the resulting metropolitan fabric as green, habitable, and *humane* as humanly possible.

References

Beatley, T. 2000. *Green urbanism: Learning from European cities.* Washington, DC: Island Press.

Brewster, G. B. 1997. *The ecology of development: Integrating the built and natural environments.* Working Paper 649. Washington, DC: Urban Land Institute.

Bullard, R. D., G. S. Johnson, and A. O. Torres, eds. 2000. *Sprawl city: Race, politics, and planning in Atlanta.* Washington, DC: Island Press.

Darling, F. F., and J. P. Milton, eds. 1965. *Future environments of North America.* Garden City, NY: Natural History Press.

Diamond, H. L., and P. F. Noonan. 1996. *Land use in America.* Washington, DC: Island Press.

Gillham, O. 2002. *The limitless city: A primer on the urban sprawl debate.* Washington, DC: Island Press.

Gottmann, J. 1961. *Megalopolis.* Cambridge: MIT Press.

Lang, R. E., and D. Dhavale. 2005a. America's megalopolitan areas. *Landlines* 17(2): 1–4.

———. 2005b. *Beyond megalopolis: Exploring America's new "megalopolitan" geography.* Census Report Series. Alexandria, VA: Metropolitan Institute at Virginia Tech.

Light, A. 2002. Restoring ecological citizenship. In *Democracy and the claims of nature,* ed. B. A. Minteer and B. P. Taylor, 153–72. New York: Rowman and Littlefield.

Lord, C., E. Strauss, and A. Toffler. 2003. Natural cities: Urban ecology and the restoration of urban ecosystems. *Virginia Environmental Law Journal* 21: 317–50.

Margulis, L., and D. Sagan. 1999. Second nature: The human primate at the borders of organism and mechanism. School of Athens Faculty Essay, reprinted in *UMass Magazine* 4(1): 24–29.

McHarg, I. 1968. *Design with nature.* Garden City, NY: Garden City Press.

Mumford, L. 1961. The natural history of urbanization. In *Man's role in changing the face of the earth,* ed. W. L. Thomas, 382–398. Chicago: University of Chicago Press.

Nicholson-Lord, D. 1987. *The greening of the cities.* London: Routledge and Kegan Paul.

Platt, R. H. 2004a. *Land use and society: Geography, law, and public policy.* Rev. ed. Washington, DC: Island Press.

————. 2004b. Toward ecological cities: Adapting to the 21st century metropolis. *Environment* 46(5): 10–27.

Platt, R. H., R. A. Rowntree, and P. C. Muick, eds. 1994. *The ecological city: Preserving and restoring urban biodiversity.* Amherst: University of Massachusetts Press.

Porter, D. R., ed. 2000. *The practice of sustainable development.* Washington, DC: Urban Land Institute.

Putnam, R. D. 2000. *Bowling alone: The collapse and revival of American community.* New York: Simon and Schuster.

Spirn, A. W. 1985. *The granite garden: Urban nature and human design.* New York: Basic Books.

Sustainable Pittsburgh. 2003. *Southwestern Pennsylvania citizens' vision for smart growth.* Pittsburgh: Sustainable Pittsburgh.

Whyte, W. H. 1957. Urban sprawl. In *The exploding metropolis: A study of the assault on urbanism and how our cities can resist it,* ed. Editors of Fortune, 115–139. New York: Doubleday Anchor.

Wilson, E. O., ed., 1988. *Biodiversity.* Washington, DC: National Academy of Sciences Press.

About the Authors

Carl Anthony is deputy director of the Conservation Resource and Development Unit at the Ford Foundation and program officer for the Foundation's Sustainable Metropolitan Communities Initiative. Before joining Ford, he was founder and executive director of Urban Habitat Program in the San Francisco Bay Area. From 1991 until 1998, he served as president of Earth Island Institute. Trained as an architect and town planner at Columbia University, he traveled in West Africa in 1970 to study how culture and natural resources shape traditional towns and villages. In 1996, he was appointed fellow at the Institute of Politics at Harvard University.

Thomas Balsley is the founder and principal designer of Thomas Balsley Associates, a New York-based landscape architecture and urban design firm specializing in the public process and design in the urban realm, with emphasis on parks, plazas, and waterfronts. His award-winning and published works include Denver's Skyline Park, Tampa City Center, Gantry Plaza State Park, Riverside Park South, Capitol Plaza, and Balsley Park in New York City, which was renamed in recognition of his contribution to the New York City public environment. Spacemaker Press has published a monograph of his work, *Thomas Balsley—The Urban Landscape.*

Timothy Beatley is the Teresa Heinz Professor of Sustainable Communities in the Department of Planning and Urban and Environmental Planning, School of Architecture, at the University of Virginia. He is author of *Ethical Land Use* (Johns Hopkins University Press, 1994) and *Habitat Conservation and Planning* (University of Texas Press, 1994).

Eugenie L. Birch is chair and professor in the Department of City and Regional Planning and codirector of the Penn Institute for Urban Research, University of Pennsylvania. She is coeditor of the University of Pennsylvania Press series The City in the 21st Century, which is reissuing Whyte's *The City: Rediscovering the Center.* She has published widely in the areas of housing and community development and planning history.

Edward J. Blakely is professor of urban planning at the University of Sydney and formerly dean and professor at the Robert J. Milano Graduate School, New School University in New York. An internationally recognized scholar in urban community development, he is coauthor of *Fortress America: Gated Communities in the United States* (The Brookings Institution, 1997).

Colin M. Cathcart, AIA, received a bachelor of environmental studies from the University of Waterloo and a master of architecture from Columbia University. Mr. Cathcart cofounded Kiss + Cathcart, Architects, which has been a leader in the sustainable design and urbanism for more than twenty years. His work has been published in the *New York Times, Metropolis, Architecture, Architectural Record, Progressive Architecture, Le Monde,* and *A Field Guide to American Architecture.* He has served on the faculties of the City College, Parsons, and Columbia, and in 1997 was appointed associate professor of Architecture at Fordham

University, where he is active in the visual arts, environmental studies, and urban studies programs.

Steven E. Clemants is a specialist on urban floras. He is vice president of the Brooklyn Botanic Garden, codirector of the Center for Urban Restoration Ecology, and president of Nature Network, a collaboration among environmental institutions in the New York metropolitan region.

Christopher A. De Sousa is an assistant professor of geography at the University of Wisconsin-Milwaukee. He is also a member of the urban studies faculty and the codirector of the newly formed Brownfields Research Consortium. He received his master of science in planning and a doctorate in geography from the University of Toronto. His research activities focus on various aspects of brownfield redevelopment in the United States and Canada. He is also active in community-based work involving urban environmental management, brownfield redevelopment, and sustainability reporting.

Steven N. Handel, a restoration ecologist of urban habitats, is interested in plant population ecology and plant-animal interactions. He received his bachelor's degree from Columbia College, and master's and doctorate in ecology and evolution from Cornell University. Previous to his Rutgers University appointment as professor of ecology and evolution, he was a biology professor at Yale University. He has been an editor of the journals *Restoration Ecology, Evolution,* and *Urban Habitats* and is an Aldo Leopold Leadership Fellow of the Ecological Society of America. He is a fellow of the American Association for the Advancement of Science, of the Australian Institute of Biology, and of the Explorers Club.

Peter Harnik is director of the Center for City Park Excellence of the Trust for Public Land. He is the author of *Inside City Parks* (Urban Land Institute, 2000) and *The Excellent City Park System: What Makes It Great and How to Get There.* A cofounder of the Rails-to-Trails Conservancy, he has also served as editor of *Environmental Action* magazine and consulted with the President's Council on Environmental Quality.

Michael C. Houck is executive director of the Urban Greenspaces Institute; urban naturalist for the Audubon Society of Portland, Oregon; and an adjunct instructor in the Geography Department at Portland State University.

Jerold S. Kayden, a lawyer and city planner, is associate professor of Urban Planning at the Harvard Design School. He is coauthor of *Privately Owned Public Space: The New York City Experience* (Wiley, 2000).

Albert LaFarge is the editor of *The Essential William H. Whyte* (Fordham University Press, 2000). He lives in Massachusetts.

Andrew Light is associate professor of philosophy and public affairs at the University of Washington. He is the coauthor of *Environment and Values* (with John O'Neill and Alan Holland, Routledge, 2006), editor or coeditor of sixteen books, including *Moral and Political Reasoning in Environmental Practice* (MIT Press, 2003), and coeditor of the journal *Ethics, Place, and Environment.*

Charles E. Little is an author and journalist specializing in land use and the environment and an adjunct faculty member (geography) at the University of New Mexico. The most recent of his fifteen books is *Sacred Lands of Indian America* (Abrams, 2003). *The Dying of the Trees* (Viking Penguin, 1995) was finalist for the Los Angeles Times Book Prize. He was formerly executive director of the Open Space Institute, director of natural resources policy at the Congressional Research Service, and president of the American Land Forum.

Anne C. Lusk is a visiting scientist at the Harvard School of Public Health and holds a Ph.D. in architecture with a focus on environment, behavior, and urban planning. She has twenty-five years of experience as a writer, researcher, and lecturer on greenways. Currently, she is conducting research on pilot products that enable physical activity through changes to the environment, research on self-identity and bike/jog/skate clothing for African American and Hispanic teen girls, and research on hospital gown designs for the Robert Wood Johnson Foundation. She won a Martin Luther King Service and Leadership Award at the University of Michigan.

Thalya Parrilla graduated from University of Massachusetts Amherst in 2004 with bachelor's degrees in anthropology and environmental sciences. She was an intern through the Environmental Careers Organization to the South Bronx in the summer of 2003, where she found her love of environmental justice work. Upon graduation, she worked with Cornell Cooperative Extension-New York City with the New Farmers Development Project and became involved with the urban agricultural scene. She now volunteers for nonprofit environmental justice organizations in the Bronx and Brooklyn.

Mary V. Rickel Pelletier collaborates with nonprofit organizations and design offices to develop innovative programs. She is currently researching improvements to stormwater management systems that can revitalize urban waterways. Recently, she coordinated the Building High-Performance Green School Programs for the Northeast Sustainable Energy Association. As visiting faculty design advisor, she assisted the Cornell University entry to the 2005 Solar Decathlon. In addition, she edited *Sustainable Architecture White Papers* (Earth Pledge Foundation, December 2000).

Rutherford H. Platt is professor of geography and planning law at the University of Massachusetts Amherst. His most recent books are *Disasters and Democracy: The Politics of Extreme Natural Events* (Island Press, 1999) and *Land Use and Society: Geography, Law, and Public Policy* (Island Press, rev. ed., 2004). He directs the Ecological Cities Project, which organized the conference from which this volume and DVD film evolved.

Deborah E. Popper teaches geography at the City University of New York's College of Staten Island, where she also participates in the environmental science, American studies, and international studies programs. She teaches regularly in the environmental studies program at Princeton University. With her husband, Frank Popper, she originated the Buffalo Commons concept for the Great Plains, which argues for ecological restoration as the organizing principle for the region's future development. She serves on the governing boards of the American Geographical Society and the Frontier Education Center.

Frank J. Popper teaches in the Bloustein School of Planning and Public Policy at Rutgers University, where he also participates in the American studies and geography departments.

He teaches regularly in the environmental studies program at Princeton University. He is author of *The President's Commissions* (Twentieth Century Fund, 1970) and *The Politics of Land-Use Reform* (University of Wisconsin Press, 1981). With his wife, Deborah Popper, he originated the Buffalo Commons concept for the Great Plains, which touched off a national debate about the future of the region. He serves on the boards of the American Planning Association, the Frontier Education Center, and the Great Plains Restoration Council.

Cynthia Rosenzweig is a senior research scientist at NASA Goddard Institute for Space Studies, where she heads the Climate Impacts Group. She was a co-leader of the Metropolitan East Coast Regional Climate Assessment of the U.S. National Assessment and is a coordinating lead author for the IPCC Working Group II Fourth Assessment Report. A Guggenheim Fellowship recipient, she has joined impact models with global and regional climate models to predict outcomes of land-based and urban systems under altered climate conditions. She holds joint appointments as professor of environmental science at Barnard College and senior research scientist at the Columbia Earth Institute.

Robert L. Ryan is an associate professor in the Department of Landscape Architecture and Regional Planning, University of Massachusetts Amherst. He is coauthor with Rachel Kaplan and Stephen Kaplan of the award-winning book *With People in Mind: Design and the Management of Everyday Nature* (Island Press, 1998). In addition, he is the codirector of the New England Greenway Vision Plan project. His current research interests include understanding the factors that affect people's attachment to parks and other open space, preserving rural character through innovative land use planning, and understanding local residents' perceptions of forest management to reduce wildland fire risk.

Laurin N. Sievert is currently a doctoral student in the Department of Geosciences at the University of Massachusetts Amherst. She earned a B.S. in natural resource management from the University of Wisconsin-Stevens Point and an M.S. in geography from the University of Massachusetts Amherst. Her current research interests include urban watershed management and stream restoration, brownfield redevelopment, and Great Lakes water supply issues.

William D. Solecki is a professor of geography and director of the Institute for New York's Nature and Environment at the City University of New York. His research focuses on urban environmental change and urban land use and suburbanization. He has served on the U.S. National Research Council, Special Committee on Problems in the Environment, and as a panel member on the NRC, Committee on Population and Land Use. He currently is a member International Human Dimensions Programme on Global Environmental Change, Urbanization and Global Environmental Change Scientific Steering Committee.

Ann Louise Strong is emeritus professor of city and regional planning at the University of Pennsylvania. Her most recent work is on land and other property restitution in the Ukraine and New Zealand.

Andrew G. Wiley-Schwartz is the director of project and program development at Project for Public Spaces. He was editor of *How to Turn a Place Around,* the acclaimed primer on community planning and revitalization, and also edited *Public Parks, Private Partners: How Partnerships are Revitalizing Urban Parks* (both published by Project for Public Spaces, 2000).

Lincoln Institute of Land Policy

The Lincoln Institute of Land Policy is a nonprofit and tax-exempt educational institution founded in 1974 to improve the quality of public debate and decisions in the areas of land policy and land-related taxation in the United States and around the world. The Institute's goals are to integrate theory and practice to better shape land policy and to provide a non-partisan forum for discussion of the multidisciplinary forces that influence public policy. This focus on land derives from the Institute's founding objective—to address the links between land policy and social and economic progress—that was identified and analyzed by Henry George.

The work of the Institute is organized in four departments: Valuation and Taxation, Planning and Urban Form, Economic and Community Development, and International Studies. We seek to inform decision making through education, research, demonstration projects, and the dissemination of information through publications, our Web site, and other media. Our programs bring together scholars, practitioners, public officials, policy advisers, and involved citizens in a collegial learning environment. The Institute does not take a particular point of view, but rather serves as a catalyst to facilitate analysis and discussion of land use and taxation issues—to make a difference today and to help policy makers plan for tomorrow.

L **LINCOLN INSTITUTE**
OF LAND POLICY
113 Brattle Street
Cambridge, MA 02138-3400 USA

Phone: 1-617-661-3016 x127 or 1-800-LAND-USE (800-526-3873)
Fax: 1-617-661-7235 or 1-800-LAND-944 (800-526-3944)
E-mail: **help@lincolninst.edu**
Web: **www.lincolninst.edu**